Finding Humanity's Origin

G.B. Templado

ISBN 0-7414-3060-6

Published by:

INFINITY
PUBLISHING.COM

1094 New DeHaven Street, Suite 100
West Conshohocken, PA 19428-2713
Info@buybooksontheweb.com
www.buybooksontheweb.com
Toll-free (877) BUY BOOK
Local Phone (610) 941-9999
Fax (610) 941-9959

Printed in the United States of America

Printed on Recycled Paper

Published May 2006

This work is dedicated to my aunt, Ms. **Eusebia L. Bryant,** whose moral support and kindness made it possible for me to write and complete this treatise.

G.B. Templado

Even if it be granted that the difference between man and his nearest allies is as great in corporeal structure as some naturalists maintain, and although we must grant that the difference between them is immense in mental power, yet the facts given declare, in the plainest manner, that man is descended from some lower form, notwithstanding that connecting links have not hitherto been discovered.

Charles Darwin
Descent of Man

Contents

The idea behind the writing of *Finding Humanity's Origin* stemmed from my search for identity. Strange to say, not only did I find my identity, but also discovered the origin of humanity as well. I found out about Charles Darwin's missing link and the Sumerian clay documents challenging Darwin's linear evolution theory that "man is descended from some lower form, notwithstanding that connecting links have not hitherto been discovered." The Sumerian documents explain why the missing link or "connecting links" will never be discovered; because there aren't any. Man was not a product of linear evolution but was created by extraterrestrial beings for purposes of their own.

The missing link is the gap in the evolutionary chain between Man and his ape ancestor. Darwin wrote in his book *Descent of Man*:

> Even if it be granted that the difference between man and his nearest allies is as great in corporeal structure as some naturalists maintain, and although we must grant that the difference between them is immense in mental power, yet the facts given declare, in the plainest manner, that man is descended from some lower form, notwithstanding that connecting links have not hitherto been discovered.

These "connecting links" are the missing parts of the puzzle that connect Homo sapiens or Man to his ape ancestor Homo erectus. These missing parts spell the difference in mental capabilities between Man and his ape forebears. Man has evolved greatly in intelligence as evinced in his abilities to build civilizations and to travel to the moon and other planets. His ape ancestor is still an ape.

Further, these "connecting links" are the heart of the Creationist vs. Evolutionist theory of human origin. As the words "Creationist" and "Evolutionist" denote, the Creationists are believers of the theory that the human race was created by God or the Supreme Being, while the Evolutionists, on the other hand, are partakers of the view that humanity came into existence through the process of linear evolution per se.

For more than a century, the battle between Creationists and Evolutionists raged on. It was a stalemate--until now. *Finding Humanity's Origin* will attempt to resolve this controversy and put an end to the battle once and for all.

The conceptual framework of this book was formulated during the thesis-writing period of my M.A. Literature program at the University of the Philippines. While I was researching my thesis entitled "Albert Camus As A Committed Writer," I came upon Charles Darwin's theory of linear evolution regarding human origin. This theory, embodied in his books, *Origin Of Species* and *Descent Of Man*, highly influenced the scientism of the 1870s in literature which was a part of the literary continuum of committed literature. In time, lincar evolution brought about the literary movements, Determinism and Naturalism, as exemplified in the works of Emile Zola and Stephen Crane. Since the evolution theory was crucial in the understanding of Albert Camus's works, I further researched the subject. I was struck by the fact that since Darwin published *Descent Of Man* in which he postulated that the "connecting links" would someday be discovered, until now these links have not been found. Sometime in the early 1900s, a substantial amount of Sumerian clay documents were unearthed by archaeologists and were later translated into English and other languages by scholars. What these clay tablets revealed is a revolutionary but plausible account of humanity's origin.

Finding Humanity's Origin is the result of my research on human origin. It is an in-depth analysis of Charles Darwin's theory of evolution and the Sumerian clay documents regarding the origin of humankind.

Since the dawn of human civilization, Man believed or was made to believe that he was the lord and master of all of God's creation on earth. Further, he also was told that the earth is the center of the universe and all planets and heavenly bodies, including the sun, revolve around it. This was the prevailing view--the religious view. In general, people paid homage to this religious doctrine because they had no scientific proof to oppose and nullify it. Before the ascendancy of science, which started with Copernicus' heliocentric finding in the 1500s, belief in an earth-centered universe with God as the Creator of the chain of being consisting of angels, man, animals, plants, and inanimate objects was the order of the day.

This view became firmly established with the appearance of Claudius Ptolemaeus or Ptolemy's book entited *The Almagest* in A.D. 137. This work, consisting of thirteen books, deals with astronomical observations and planetary movements. Its central idea is that all heavenly bodies move in a fixed pattern and predictable manner, and that the earth is the center of the universe. This book found favor with the Church fathers, and they lauded Ptolemy for this achievement. The Church accepted Ptolemy's scheme of things for two reasons: first, it fostered its dogma of a divinely ordered universe where Man is lord and master of the universe since planet earth is the center of the universe, and all heavenly bodies follow a set of rules just the way God wanted it to be; and second, it made it easier for the Church to fulfill its function which is to govern and control its flock. It attempted to control people's minds by preying on their egos and self-esteem. And it worked because it satisfied a basic human mental/psychological need--the need to feel important. If Man wanted to keep his exalted and important position in the chain of being, all he has to do is obey the teachings of the Church, God's representative on earth. A write-up from

Collier's Encyclopedia commented on the Church's interpretation of Ptolemy's view of the universe:

> Ptolemy's systematic ordering of the planets . . . was interpreted by the Church as a part of a great hierarchy or chain of being with God and the angels as the highest link, then man, woman, animals, plants, and stones as the lower links of the chain.[1]

This view on the chain of being held strong and prevailed until the end of the 1400s. Its foundation was shaken by Nicolaus Copernicus' scientific findings in the 1500s. In his book, *On the Revolution of Heavenly Bodies*, published in 1543, Copernicus stated that the sun is the center of our solar system, and that the earth is just one of the planets that revolves around the sun. By using more sophisticated instruments than Ptolemy, he (Copernicus) was able to plot out accurate measurement of planetary movements and approximated earth's rate of motion. His heliocentric finding was accepted by astronomers who came after him because they were able to produce the same results. The following is a comment on the acceptance of Copernicus' theory:

> The theory secured general acceptance among leading astronomers within a century and a half of its formulation. It was presupposed in Kepler's enunciation of the fundamental laws of planetary motion and Newton in 1687 interpreted these laws as necessary consequences of a gra-

---

[1] *Collier's Encyclopedia*, vol. 19, 1973, 480.

> vitational attraction of the planets towards the sun.[2]

Needless to say, Copernicus' heliocentric finding, later confirmed by Galileo, impacted Man's self-concept, not to mention its devastating effect on the Church. It diminished Man's feeling of self-importance and made him realize that the planet he inhabits is not the center of the universe as he thought but is just one of the planets that revolves around the sun. To add insult to injury, astronomers also found out that our solar system is just one of the numerous solar systems in the universe and an insignificant one at that, making Man merely a speck in the scheme of things. If that were not enough, Man's significance was further reduced by the advent of more scientific findings and theories inimical to his position in the chain of being.

Perhaps, equal to or greater than Copernicus' heliocentric finding, which reduced Man's importance in the chain of being, is Charles Darwin's theory of linear evolution. In his two books *Origin of Species* and *Descent of Man*, Darwin postulated that Man is merely a higher animal, a product of linear evolution and not specially created by God or the Supreme Being.

Darwin believed that all forms of life on earth, including Man, was a product of evolution through the process of natural selection and not created by God. Nature selects what species are to survive and what species are to perish. He wrote in the *Origin of Species:*

> Natural selection is daily and hourly scrutinizing throughout the world, every

---

[2] *Collier's Encyclopedia*, vol. 7, 1973, 303.

> variation, even the slightest; rejecting that which is bad, preserving and adding up all that is good; silently and insensibly working, whenever and wherever opportunity offers, at the improvement of each organic being in relation to its organic and inorganic conditions of life. We see nothing of these slow changes in progress, until the hand of time has marked the long lapse of ages, that we only see that the forms of life are now different from what they formerly were.[3]

Charles Darwin's theory of natural selection did not spring suddenly from his head but took a while to germinate. In fact, the incubation period of this theory took a very long time. (Fig. 1)

Long before Darwin became a professional naturalist, he had been an amateur collector of sorts. His "passion for collecting" manifested itself early in life. He recalled that by the time he attended a day school in Shrewsbury at the age of eight, his expertise at collecting had been developed. He wrote in his autobiography which was reprinted in Robert Jastrow's book entitled *The Essential Darwin*:

> By the time I went to this day-school [Shrewsbury] my taste for natural history, and more especially for collecting, was well-developed. I tried to make out the names of plants, and collected all sorts of things, shells, seals, corns, and minerals. The passion for collecting, which leads a man to be a systematic naturalist, a virtuoso or a miser, was very

---

[3] Charles Darwin, *Origin of Species*, 3.

Courtesy of National Portrait Gallery, London

Charles Darwin
Portrait by John Collier (Fig. 1)

> strong in me, and was clearly innate, as none of my sisters or brother ever had this taste.[4]

This passion for collecting continued through the years, well before he chose the profession or career of naturalist.

Before Darwin chose the career of naturalist, he tried two other professions, medicine and theology. He first matriculated at the medical school in Edinburgh but finding his subjects dull and boring, he was not inclined to do well in this course. Needless to say, since his subjects did not interest him, he did not apply himself, and his grades were barely satisfactory. His unsatisfactory performance prompted his father, Dr. Robert Darwin, into thinking that his son was not suited to be a physician and pulled him out from medical school. For the young Darwin's part, he hated his academic program, saying:

> The instruction at Edinburgh was altogether by Lectures, and these were intolerably dull, with the exception of those on chemistry by Hope; but to my mind there are no advantages and many disadvantages in lectures compared with reading. Dr. Duncan's lectures on Materia Medica at 8 o'clock on a winter's morning are something fearful to remember. Dr. Munro made his lectures on human anatomy as dull as he was himself, and the subject disgusted me.[5]

---

[4] Robert Jastrow, *The Essential Darwin*, 3.
[5] Ibid., 5.

Since Charles could not be a physician, the older Darwin suggested to his son that he pursue the study of theology. He had no objection to this suggestion, but deep down the young Darwin knew that he wasn't cut out to be a cleric of the Church of England. For one thing, his hyper-analytical mind was strongly skeptical about dogma, and that's what the Bible and Scriptures are all about--dogmatic treatises on the history of humankind, including its creation. And another thing, the academic regimentation of the school of divinity was constricting. Although he had no intention of rejecting the teachings of the Church, he didn't blindly swallow the dogmas either. He opted to reserve judgment until the Creeds became intelligible to him. He wrote in his autobiography:

> He [Dr. Robert Darwin] proposed that I should become a clergyman . . . I asked for some time to consider as from what little I had heard and thought on the subject I had scruples about declaring my belief in all the dogmas of the Church of England . . . I read with care Pearson on the Creeds and a few other books on divinity; and as I did not then in the least doubt the strict and literal truth of every word in the Bible, I soon persuaded myself that our Creed must be fully accepted. It never struck me how illogical it was to say that I believed in what I could not understand and what is in fact unintelligible. I might have said with entire truth that I had no wish to dispute any dogma; but I never was such a fool and say "credo quia incredible."[6]

---

[6] Gavin de Beer, *Autobiographies: Charles Darwin, Thomas Henry Huxley*, 31.

Like his medical training, he likewise did poorly in his program in divinity. But before he could be academically dismissed from the school of divinity at Cambridge, an opportunity to serve as naturalist on the *Beagle* opened up. Darwin enthusiastically signed up for the position of naturalist. At first his father was against his son's decision but was soon prevailed upon by an uncle who reasoned that if the young Darwin did badly in his medical and theological training, it was because he didn't like both disciplines. These two academic programs were imposed on him; he did not choose either. "Why not give him what he wants?" asked the uncle.

Dr. Robert Darwin was a reasonable man. Having seen that the uncle made sense, he gave his permission for his son to accept the position of naturalist.

The purpose of the voyage of the *Beagle* was to circumnavigate the world and charter islands that were part of England's sphere of influence. Aboard the *Beagle*, and without the constraints and demands of everyday living, Darwin pondered seriously the theory of natural selection.

The concept of linear evolution by natural selection, that is, the uninterrupted development of life forms or species without supernatural or divine intervention, had been in the air for quite sometime. We find this concept illustrated in the works of famous scientists of the nineteenth century, namely, Jean Baptiste Lamarck, Alfred Wallace, etc. In fact, Wallace and Darwin came up with the natural selection theory independently. Officially though, Wallace had priority over the natural selection theory because there is documentation that Wallace published an essay on natural selection in 1858. Darwin's *Origin of Species* was not published until November 1859.

In June of 1858, Alfred Wallace sent an essay on natural selection to Darwin from the East Indies. Wallace had been collecting specimens there for some time. Wallace wrote:

> It is the object of the present paper to show . . . that there is a general principle in nature which will cause many varieties to survive the parent species, and to give rise to successive variations departing further and further from the original type.[7]

Darwin was startled that anyone had arrived at a similar conclusion on the evolution of species. He commented dryly:

> I never saw a striking coincidence . . . If Wallace had my MS sketch written out in 1842 he could not have made a better short abstract.[8]

Attached to Wallace's essay was a note requesting Darwin to forward this study to Sir Charles Lyell, a noted geologist of the time. Lyell had established his reputation as a scientist with his book *Principles of Geology* and was respected by the scientific community. Wallace thought that a favourable word from Lyell would bolster his career as a naturalist. Wallace at this time was an emerging scientist of sorts.

Darwin was in a quandary. What should he do? Wallace had destroyed the originality of the conceptual framework of the *Origin of Species*. Darwin was not the originator

---

[7] Janet Browne, *Charles Darwin: The Power of Place*, 14.
[8] Ibid., 15.

of the theory of natural selection as he had thought. Several options raced through his mind as he pondered on the situation. Should he discredit Wallace and claim priority to the natural selection concept or uphold the truth? It was a battle between his reputation as a scientist versus his conscience. After an intense inner struggle, he decided to uphold the truth. He forwarded the essay to Lyell as Wallace had requested. Darwin wrote:

> Some years ago, you recommended me to read a paper by Wallace in the *Annals* which interested you and as I was writing to him, I knew this would please him much, so I told him. He has today sent me the enclosed and asked me to forward it to you. It seems to me well worth reading. Your words have come true with a vengeance that I should be forestalled. You said this when I explained to you here very briefly my views of Natural Selection depending on the struggle for existence . . . Please return me the MS which he does not say he wishes me to publish, but I shall of course at once write and offer to send to any Journal. So all my originality, whatever it may amount to, will be smashed. Though my Book, if it will ever have any value, will not be deteriorated, as all the labour consists in the application of the theory. I hope you will approve of Wallace's sketch, that I may tell him what to say.[9]

---

[9] Ibid., 17

The issue on authorship of the evolution theory by natural selection did not erupt into a conflict. Wallace conceded to Darwin as having priority of authorship. Wallace made this decision after having talked to his and Darwin's scientist friends.They confirmed that Darwin had been studying this theory long before Wallace published his eassay. The following is a letter from Alfred Wallace to Charles Kingsley on May 7, 1869. Wallace wrote:

> As to C. Darwin, I know exactly our relative positions, and my great inferiority to him. I compare myself to Guerilla Chief, very well for a skirmish or for a flank movement, and even able to sketch out the plan of campaign, but reckless of communications and careless about Commissariat--while Darwin is the great General, who can maneuver the largest army, and by attending to his lines of communication with an impregnable base of operations, and forgetting no detail of discipline, arms or supplies, leads on his forces to victory. I feel truly thankful that Darwin had been studying the subject [natural selection] so many years before me, and that I was not left to attempt and to fail, in the great work he has so admirably performed.[10]

True to Wallace's statement that Darwin "had been studying the subject" of natural selection for a long time before him, Darwin, indeed, had been working on the evolution theory by natural selection in the early 1830s before his return to England in 1836 from his trip aboard the *Beagle*

---

[10] Ibid., 317.

where he served as naturalist. As naturalist, his job was to collect flora and fauna from different parts of the world, catalogue and classify them. At first, collecting specimens was just a task for him, a job he had to do as payment for his room and board. In point of fact, it was no job at all for he had been collecting living things since he was a child. The task of naturalist was nothing but a resumption of his childhood hobby. But as he got more involved with the data-gathering it struck him that he actually enjoyed the process of collecting living things as he used to in his childhood days. Slowly and gradually, he came to the realization that he had found his true calling, and that he wanted to study living things not just as a hobby during his spare time but make it his lifework, his lifelong career. He wanted to study and determine the whys and wherefores of living things' existence. Thereupon, he embarked on the greatest adventure of his life, that is, observing plants and animals in their natural habitat, and writing down the results of his observations and experiments as to how these living matter respond to different stimuli and evolved into new varieties and species.

Upon his return to England in 1836, he organized his notes into coherent data. He pondered and tried to formulate a conceptual framework that would unify all the information. It took a lot of work and time to research different scientific books, magazines and journals. Finally, in October 1838, he found what he had been looking for in Malthus's essay on population. Briefly, Malthus's theory on population states that human population tends to grow geometrically while food production tends to increase arithmetically; in the course of time, there will be scarcity of food supply because food production will not be able to meet the needs of all humans; in consequence, there will be a fierce struggle for existence, where only the fittest will survive. This theory, which Darwin called natural selection, was just what he needed for the unifying theme of his voluminous work. Darwin wrote in his autobiography:

> In October 1838, that is, fifteen months after I had begun my systematic enquiry, I happened to read for amusement *Malthus on Population*, and being well prepared to appreciate the struggle for existence which everywhere goes on from long-continued observation of the habits of animals and plants, it at once struck me that under these circumstances favourable variations would tend to be preserved, and unfavourable ones to be destroyed. The result of this would be the formation of new species. Here, then, I had at last got a theory by which to work.[11]

One other notable naturalist who came up with the natural selection theory was a Frenchman named Jean Baptiste Lamarck. Unlike Wallace and Darwin who merely stated that natural selection is the driving force in the evolution of species, Lamarck, on the other hand, went further by stipulating in detail that organisms possess an inner force which pushes them forward or holds them backward as they interact with their environment. Lamarck further believed that nature's plan is development of all organic forms toward complexity, but a "multitude of circumstances" in the organisms' environment destroy the "regularity in the gradation of the growing complexity of organization." How organisms respond to the stimuli in their environment brings about organic change. The body part of an organism that is not in use will eventually be shed off and disappear. A part that is constantly used by an organism in its struggle for survival will be-

---

[11] Nora Barlow, *The Autobiography of Charles Darwin1809-1882*, 120.

come stronger and will be transmitted to its offspring. In biology, this phenomenon is referred to as the Use and Disuse Doctrine. In time, the offspring will evolve and metamorphose into a new and distinct species with inherited characteristics. In his book, *Philosophie Zoology*, published in 1809, Lamarck illustrated the Use and Disuse concept in two laws. The following is the reprint and translation of these laws:

> *First law*: In every animal that has not reached the end of its development, the more frequent and sustained use of any organ will strengthen this organ little by little, develop it, enlarge it, and give to it a power proportionate to the duration of this use; while the constant disuse of such an organ will insensibly weaken it, deteriorate it, progressively diminish its faculties, and finally cause it to disappear.
>
> *Second law*: All that nature has caused individuals to gain or lose by the influence of the circumstances to which their race has been exposed for a long time, and, consequently, by the influence of a predominant use or constant disuse of an organ or part, is conserved through generation in the new individuals descending from them, provided that these acquired changes are common to the two sexes or to those which have produced these new individuals.[12]

---

[12] Richard W. Burkhardt, Jr., *The Spirit of System: Lamarck and Evolutionary Biology*, 166.

This Use and Disuse concept was later amplified by Darwin in the chapter on rudiments. In Chapter I of his book *Descent of Man*, he discussed in detail these rudiments or inherited characteristics to support his theory that Man descended from the lower forms: that Man inherited rudiments from the lower organic forms but later shed them off because he no longer needed these rudiments or physical characteristics for his survival.

For a while, the theory of linear evolution by natural selection was just another theory and didn't have a firm footing in the scientific world. It wasn't until the publication of the *Origin of Species* in 1859 that the theory became established in the annals of science. The reason for this was that Darwin was able to support, by voluminous documentation, the theory that plants and animals evolve through the process called natural selection per se, that is, nature selects which species are to perish and which species are to survive and form other distinct species. God, therefore, did not play a part in the existence of life forms, either plant or animal. Since this theory was controversial and flatly denied the prevailing Creationist theory that God created the world and all living things on earth, Darwin knew that the *Origin of Species* would be attacked by scientists and laymen alike. So, before he published the book, he took some measures to protect himself from the imminent onslaught--he hired a champion.

"Why didn't Darwin defend his views himself?" one may ask. He didn't because he got physically sick when talking before an audience. In fact, he was chronically ill after his return from the Galapagos Islands (where he collected most of his specimens) until his death in 1882. Therefore, he sought the services of a scientist who would champion his theory. He thought that Thomas Huxley was the best man for the job. Huxley was a university lecturer by profession, a published author, and had a knack for oratorical debate.

Several months before the publication of the *Origin of Species*, Darwin requested Huxley's support and tried to convince him that he would be fulfilling his true function as a scientist if he championed the evolution doctrine. In 1859, the same year the book was published, he wrote, "If I can convert Huxley, I shall be content."[13]

At first, Huxley denied the request because, like most men of his generation, he was brought up with the Creation story, and the Book of Genesis was ingrained in his mind. A struggle between this indoctrination and his training as a scientist ensued. After a lot of soul-searching and reflection, the scientist in him won. Darwin's argument and the voluminous data in the *Origin of Species* outweighed the Creation tale. Huxley wrote of this conversion:

> My reflection, when I first made myself master of the central idea of the *Origin* was "how extremely stupid not to have thought of that." I suppose that Columbus' companions said much the same when he made the egg stand on end. The facts of variability, of the struggle of existence, adaptations to conditions, were notorious enough, but none of us had suspected that the road to the heart of the species problem lay through them, until Darwin and Wallace dispelled the darkness and the beacon-fire of the *Origin* guided the benighted.[14]

[13] Don Nardo, *Charles Darwin*, 128.
[14] Ibid., 131.

Defending Darwin's ape/man theory was no bed of roses--this Huxley found out in the course of time. True to Darwin's expectations, the *Origin of Species* was attacked by the scientific and non-scientific sectors of society. Huxley, as his champion, became a target of scurrilous remarks in both the press and media. In several instances, Huxley found himself an object of ridicule. A case in point was the Oxford Debate of June 28, 1860 of Darwin's theory. This debate was in the form of an oratorical competition consisting of two parties. One party argued FOR and the other party argued AGAINST the theory. The FOR side was represented by Huxley, coached by his client Charles Darwin; the AGAINST side was headed by the Bishop of Oxford, tutored by Professor Richard Owen, the most eminent paleontologist and comparative anatomist of the time. Several newspapers, journals, and magazines recorded this debate. Perhaps the best of these reports was that of Francis Darwin, Charles Darwin's son, because it captured the spirit or essence of the debate. The following is the highlight of the debate as reported by Francis Darwin:

> The Bishop of Oxford attacked Darwin, at first playfully but at last in grim earnest. It was known that the Bishop had written an article against Darwin in the last *Quarterly Review*. It was also rumored that Prof. Richard Owen had been staying at Cuddesden and had primed the Bishop, who was to act as mouthpiece to the great Paleontologist, who did not dare to enter the list [contest] . . . The Bishop was declaring with rhetorical exaggeration that there was hardly any actual evidence on Darwin's side . . . "I should like to ask Professor Huxley, who is sitting by me, and is about to tear

> me to pieces when I have sat down, as to his belief in being descended from an ape, is it on his grandfather's or his grandmother's side that the ape ancestry comes in?"[15]

Although the debate ended amicably since all the participants were professionals, everyone present at the meeting knew that the issue was far from being resolved. In fact, it was just the beginning. Future events proved this prediction to be right.

Darwin's theory of linear evolution of plants and animals, by natural selection including Man, is contained in his two seminal books, namely, *Origin of Species* and *Descent of Man*. Whereas in the *Origin of Species* he merely alluded that Man evolved from the lower organic form, in the *Descent of Man* he made an unequivocal statement that Man descended from the lower forms or animals. He wrote:

> He who wishes to decide whether man is the modified descendant of pre-existing form, would probably first enquire whether man varies, however slightly, in bodily structure and in mental faculties; and if so, whether the variations are transmitted to his offspring in accordance with the laws which prevail with the lower animals . . . It might also naturally be enquired whether man, like so many other animals, has given rise to varieties, and subraces . . . We shall see that all of these questions, as indeed is obvious in respect to most

[15] Francis Darwin, *"The Famous Oxford Debate of Darwin's Theory"*, 150.

> of them, must be answered in the affirmative, in the same manner as with the lower animals.[16]

The *Descent of Man* discusses mainly Man's origin--how he evolved from the lower organic forms; the *Origin of Species* deals mainly with plants and a few birds and insects--how they evolved into different varieties and species. But to most readers, both books are the same, i.e., they both deal with Darwin's ape/man theory. So, when a respondent comments or an author writes about the *Origin of Species* or *Descent of Man*, what he or she is talking about, most of the time, is solely Darwin's linear evolution theory that Man descended from the ape.

[16] Charles Darwin, *Descent of Man*, 5-6.

Reaction to the linear evolution theory which states that Man evolved from the ape was three-fold. The first group subscribed to the theory, the second was neutral, and the third was against. The prime exponent of the first group was, of course, Thomas Huxley. Not only did he defend Darwin's theory publicly but did so in writing as well. In a series of public lectures that were delivered at the School of Mines in London, he disseminated Darwin's ape/man theory to the working people of the city. It was a sight to behold—individuals of the lower strata of society with minimal education and meager income flocking to hear Huxley explain Man's origin and forebears. In his flamboyant style, Huxley was able to enthrall his audience with a plethora of scientific data borrowed from his master Charles Darwin. As he was able to captivate and mesmerize his listeners, the popularity of his lectures spread like wildfire and people from the outskirts of the city came to listen to him. To date, these lectures are the most well-attended public academic meetings in history since Socrates' mobile school for the young men of Athens thousands of years ago. In a way, the lectures were a significant victory for the proponents of linear evolution theory in the sense that a scientific theory captivated and sustained the interest of the lower ranks of a society. In a letter to his wife, Huxley wrote:

> My working men stick by me wonderfully the house being fuller than ever last night. By next Friday evening they will all be convinced they are all monkeys.[17]

Likewise, in several essays and books Huxley published on Darwin's ape/man theory, he courageously and

---

[17] Leonard Huxley, *The Life and Letters of T.H. Huxley*, 190.

unwaveringly defended this theory. In his book, *Man's Place in Nature*, Huxley wrote:

> I have endeavored to show that no absolute structural line of demarcation wider than that between the animal which immediately succeed us in the scale can be drawn between the animal world and ourselves . . . the attempt to draw a psychical distinction is equally futile, and that even the highest faculties of feeling and of intellect begin to germinate in lower forms of life.[18]

One of the influential political scientists of the time who was impressed by Darwin's work was Karl Marx, whose book entitled *Das Capital* became the bible of Communism, a totalitarian form of political organization. Karl Marx sent Darwin a copy of *Das Capital* and requested his comments. The following is Darwin's response:

> I heartily wish I was more worthy to receive it, by understanding more of the deep and important subject of political economy. Though our studies have been so different, I believe that we both earnestly desire the extension of knowledge and that this in the long run is sure to add to the happiness of mankind.[19]

The social scientist who spoke favourably of Darwin's work was John Stuart Mill, the father of Utilitarian-

---

[18] Thomas Henry Huxley, *Evidence as to Man's Place in Nature*, 109.
[19] Janet Browne, *Charles Darwin: The Power of Place*, 403.

ism, a philosophy which states that the greatest good is that which benefits the greatest number, which influenced greatly the government policies of democratic societies like the United States of America. He defended Darwin's line of reasoning in the *Origin of Species*, saying:

> Mr. Darwin's remarkable speculation on the *Origin of Species* is another unimpeachable example of a legitimate hypothesis . . . It is reasonable to accuse Mr. Darwin (as has been done) of violating the rules of induction. The rules of induction are concerned with the condition of proof. Mr. Darwin has never pretended that his doctrine was proved. He was not bound by the rules of induction but by those of hypothesis. And these laws have seldom been more completely fulfilled. He has opened a path of inquiry full of promise, the results of which none can foresee.[20]

Two notable theologians who were in favor of Darwin's evolution theory were Charles Kingsley and Baden Powell.

Kingsley was both a theologian and naturalist. Unlike his contemporary, the Rev. Sedgwick who threatened hellfire to adherents of evolution/natural selection, Kingsley, praised Darwin's book. Janet Brown, in her biography of Darwin, entitled, *Charles Darwin: The Power of Place*, said of Kingsley:

---

[20] John Stuart Mill, *A System of Logic, Ratiocinative and Inductive*, 18.

> He [Kingsley] appeared sincerely interested in Darwin's arguments, telling Darwin he could imagine an all-wise, all-powerful deity making organisms that make themselves. Indeed, he was the first clergyman to see in Darwin's schemes an internal beauty that could be shared by science and spiritual revelation alike. To Kingsley, scientists who tried to explain away the existence of the spirit would be as intellectually stunted as clerics who attempted to dismiss the truths of science.[21]

The other theologian who spoke highly of Darwin's work was Rev. Baden Powell. Unfortunately for Rev. Powell, he was under the clerical jurisdiction of Samuel Wilberforce, the Bishop of Oxford. If the reader recalls, the Bishop of Oxford was the mouthpiece of Richard Owen in the Oxford Debate of 1860 regarding Darwin's theory. Richard Owen, the most eminent comparative anatomist at the time, challenged Darwin's natural selection theory. Upon knowing that a member of his flock had strayed from the righteous path, the Bishop initiated a campaign against Powell who considered the *Origin of Species* a great book, " a work which must soon bring about an entire revolution of opinion in favor of the grand principle of the self-evolving powers of nature."[22] Powell died before his treatise supporting natural selection was published. If he had lived, he certainly would have been a target of the Bishop's ecclesiastical shotgun.

The second or neutral group consisted of three factions, namely, the Queen of England, the scientist, and the lay

---

[21] Janet Browne, *Charles Darwin: The Power of Place*, 95.
[22] Ibid., 114.

person. All three factions did not criticize or subscribe to the theory of natural selection. Except for Queen Victoria who took a non-committal position, the two groups viewed natural selection as an important addition to knowledge and made it known to the public that Darwin's effort in postulating the idea was salutatory.

Sometime in the latter part of 1859, Charles Lyell and some distinguished intellectuals (philosophers, literary savants and scientists) met with the prince consort in a public lecture meeting hosted by the British Association. This was an academic gathering of sorts. The meeting was an intellectual discussion--an exchange of literary, philosophical, and scientific ideas. As the meeting progressed, the main discussion centered and focused on the most interesting idea of the day, Darwin's theory of linear evolution by natural selection. All three groups of intellectuals realized that the theory would have a profound impact in their respective fields should Darwin's theory prove to be right. It was a momentous event comparable to Copernicus' heliocentric finding. If science could prove that Darwin was right, all of humankind's knowledge would need to be re-evaluated and changed, just as the heliocentric view altered human learning and institutions.

Later on, Lyell enthusiastically recapitulated this meeting to his family, telling them how the participants, including the prince consort, were impressed with Darwin's theory and how he (Lyell) stood out as the scholar of the day. Although Queen Victoria merely mentioned this meeting in her diary as a day "laden with philosophers and savants,"[23] Darwin's theory of evolution did capture her interest for she sent the prince consort to attend this meeting and asked him

---

[23] Elizabeth Longford, *Victoria R.I.*, 286.

to brief her about the gathering.

The main proponent of the scientists who belonged to the second or neutral group was William B. Carpenter, a contemporary of Darwin. He commented on natural selection, saying:

> The doctrine of progressive modification by Natural Selection propounded by Mr. Darwin enables us to look at the War of Nature constantly going on around us as not marked only by suffering and death but as inevitably tending towards the progressive exaltation of the races engaged in it; just as, in the world of mind; it is only by intellectual collision that Truth can become firmly established and only by moral conflict, whether in the individual or in society; that Right can obtain an undisputed sway.[24]

The "man in the street" or lay person who belonged to the neutral group was George Harris, a tailor. Like the naturalist Carpenter who believed that the *Origin of Species* broadened one's horizon's and contributed to our understanding of nature, Harris similarly viewed the book as a significant addition to knowledge. Harris wrote to Darwin on March 3, 1862, requesting a copy of the *Origin of Species*:

> Being but a poor working man with a wife and four children to support, and further, being one of those thirsty souls who

[24] David Hull, *Darwin and His Critics*, 114.

> would drink deep at the fountain of Science --when accessible, I have taken the liberty to obtrude upon you and to ask of you the favor of a presentation copy of your great work on the *Origin of Species* the price being far above my purchasing means. As I am but a poor tailor, I am desirous not to ask the labor of others for nothing, and in return, be but too glad to do work for the amount. "My soul thirsteth after Knowledge" is the apology I make for my obtrusion.[25]

The third group who was against Darwin's ape/man theory was comprised of the "man in the street," clergymen, and scientists.

The popular or "man in the street" argument against Darwin's linear evolution theory is that if apes changed into humans long ago, why don't they do so now? Wiktor Stoczkowski, in his book *Explaining Human Origins*, sums up this "man in the street" question:

> . . . frequently raised [question] by the man in the street: if long ago . . . an ape actually changed into a human being, why do apes no longer transform themselves into humans today?[26]

Likewise, several clergymen were against the *Origin of Species*. The spokesman of this group was Adam Sedgwick, Darwin's former teacher. Rev. Sedgwick was both a

---

[25] *The Correspondence of Charles Darwin*, vol. 10, 99.
[26] Wiktor Stoczkowski, *Explaining Human Origins*, 61.

scientist and theologian. He believed that the scientist should assist the theologian in accomplishing his goal, that is, science should serve God by showing humanity the righteous path or the way to God. Sedgwick thought that he was the warden of all souls, including Darwin's. For this reason, he embarked on a personal crusade, both orally and in writing, criticizing the *Origin of Species* as an instrument of evil because the book rejected God as the creator of all living things including humankind. In one of his sermons, he reminded his flock that God will consign anyone to hell who denies Him as The Maker. Also, in his philosophical treatise entitled *Discourse on the Studies of Cambridge*, he stated in the preface that "evolution will bring about spiritual chaos." This crusade, Rev. Sedgwick thought, will bring Darwin to his senses and he will recant. He was wrong. Darwin stuck to his viewpoint through the end of his life.

In his time, Sedgwick embodied the prevailing view about life on earth. Up to the late eighteenth century, the dominant view about the world and Man's place in it was theological or religious: that God or the Supreme Being created all things on earth, plants and animals, inluding Man. But as time passed and as we came to know the earth better, the religious view slowly and gradually became suspect. One other historical event that struck a major blow of this view following Copernicus' heliocentric finding and Darwin's evolution theory was the discovery of animal fossils during the Industrial Revolution.

As the Industrial Revolution gained momentum, there was more demand for sources of energy, particularly coal, to fuel newly invented machines to produce goods and services. To fill this need, there was extensive digging on the surface and excavating underneath the earth for coal. As coal was processed into usable fuel, some workers noticed that along with the coal were bones or fossils of animals. The workers brought these fossils to their supervisors, and the supervisors, thinking that these fossils may have scientific value, took them to the authorities for analysis. Results of the examination of these fossils revealed that they belonged to a group of animals that no longer existed which contradicted the Bible. Roger Lewin's book, *The Origin of Modern Humans*, commented on this account:

> In the late eighteenth century the digging of canals and the excavation of coal mines exposed successive sedimentary layers of rock and the fossils they contained. Increasingly it became apparent that some of the fossilized bones belonged to species that no longer existed; in some cases a progression of forms could be traced from lower to higher sedimentary layers. Since Noah was supposed to have rescued all living species, these discoveries revealed that biblical account must be incomplete.[27]

Here, we have another major clash between science and religion. The seed of doubt that was planted in people's minds with Copernicus' finding that the earth is not the center of the universe took root with Darwin's evolution theory and germinated with the discovery of the fossils. These factors eroded a lot of the authority and credibility of the religious view and changed Man's mental set forever. Andre Gide, a Nobel Prize laureate, best describes this situation when he said that once an individual's mind is stretched to a new idea it will never go back to its original dimension.

---

[27] Roger Lewin, *The Origin of Modern Humans*, 40.

The leading scientists who opposed Darwin's linear evolution theory regarding Man were Richard Milner, William Whewell, Alfred Wallace, and Robert Foley.

Milner's objection to Darwin's theory of linear evolution regarding Man's descent from apes was that: if Man descended from the ape, Homo erectus, where were the intermediate species or "connecting links" that separate Homo erectus from Homo sapiens or Man? Milner wanted to know. In his book *Charles Darwin: Evolution of a Naturalist*, Milner asked his fellow naturalists the following question: "If evolution was really a fact, where are the missing links?" Milner was of course referring to the "connecting links" or several species that connect Homo erectus to Homo sapiens because the gap in mental powers between ape and Man is immense. The "connecting links", if there were any, would certainly show how instinct in the ape developed into intelligence in Man. Why hadn't these links been found?

Darwin responded to this question saying that the reason why the "connecting links" had not been found was because of a lack of reliable geological equipment. He anticipated that the process of finding the fossils of the "connecting links" or intermediate species would take a long time due to the poor equipment being used. Geologists and/or paleontologists had not yet developed a sophisticated machine to unearth these fossils. In consequence, the geological record was imperfect or found wanting. In his book *Origin of Species*, Darwin discussed this geological problem:

> On this doctrine of the extermination of an infinitude of connecting links, between the living and extinct inhabitants of the world and at each successive period between the extinct and still older species why is it not every geo-

> logical formation charged with such links? Why does not every collection of fossil remains afford plain evidence of the gradation and mutation of the forms of life? We meet with no such evidence, and this is the most obvious and forcible of the many objections which may be urged against my theory. Why do we not find great piles of strata beneath the Silurian groups of fossils? For certainly on my theory such strata must somewhere have been deposited at these ancient and utterly unknown epochs in the world's history.
>
> I can answer these questions and grave objections that the geological record is far more imperfect than most geologists believe . . . only a small portion of the world has been geologically explored.[28]

With regards to the gap in mental powers between Homo erectus and Homo sapiens or Man, Darwin agreed that it is true that the difference in the mental capabilities between ape and Man is immense. But it is also true that there is no fundamental difference in their brains--that the difference between the two brains are "in degree and not of kind." The higher faculties were gradually developed from ape to Man over a long period of time. The "connecting links" or missing links, Darwin believed, would prove this. He wrote in the *Descent of Man*:

> Even if it be granted that the difference between man and his nearest allies is as great in corporeal structure as

---

[28] Charles Darwin, *Origin of Species*, 374-75.

> some naturalist maintain, and although we must grant that the difference between them is immense in mental power, yet the facts given declare, in the plainest manner, that man is descended from some lower form, notwithstanding that connecting links have not hitherto been discovered.[29]

Perhaps, the most notable opponent to the theory that Man was a product of linear evolution was Alfred Wallace. Like Milner, Wallace believed that Man couldn't have e-volved from the ape because of the tremendous difference in mental capabilities of the two species. While he conceded to Darwin that plants and the lower organic forms evolved by natural selection, he believed that Man did not because of his mental powers. He argued further that Man couldn't have de-veloped these mental capabilities in the span of time that he branched out from our immediate ancestor Homo erectus. Wallace believed that an external force or agent had inter-vened in the evolution of Homo erectus. He said that:

> At some point in mankind's early history, physical evolution stopped and some higher driving force or spirit had ta-ken over. The human mind alone continued to advance, human societies emerged, cul-tural imperatives took over, a mental and moral nature became significant and civili-zation took shape. Modern mankind thus escaped the fierce scrutiny of natural select-ion. The development of human thought freed humanity from the inexorable laws of nature . . . Mankind was composed of a ma-

[29] Charles Darwin, *Descent of Man*, 151.

> terial [descended from the apes] and an immaterial spirit (infused by a higher power) that pulled mental and cultural development onwards.[30]

Wallace's argument against Man's evolution by natural selection that Man couldn't have been a product of linear evolution because his mental qualities could not have evolved by "mere natural selection" was supported by several eminent paleontologists. Perhaps, the most famous of these paleontologists was Richard Leakey who commented on Wallace's viewpoint. He said:

> The nineteenth century English naturalist Alfred Russell Wallace--who also invented the theory of natural selection, independently of Darwin--balked at applying the theory to those aspects of humanity we most value. He considered humans too intelligent, too refined, too sophisticated to have been the product of mere natural selection. Primitive hunter-gatherers would have had no biological need for these qualities, he reasoned, and so they could not have arisen by natural selection. Supernatural intervention, he felt, must have occurred to make humans so special.[31]

Darwin responded to Wallace's argument against Man as a product of linear evolution by giving a lengthy and detailed account explaining why Man or humans are "so spe-

---

[30] Janet Browne, *Charles Darwin: The Power of Place*, 318.

[31] Richard Leakey, *The Origin of Humankind*, 4.

cial." Darwin averred that if humans are different from the apes it's not because of "supernatural intervention" or "higher driving force or spirit," as Wallace inferred, but because in the course of Homo erectus' evolution the forces of nature became more fierce and demanding. Food became scarce. The intermediate species or what Darwin called "connecting links" had many enemies and rivals in the procurement of food. In order to survive, the species that evolved to Homo sapiens or Man (let's call him Species X) had to "think" and devise ways and means to defeat his enemies to obtain food. In the process of "thinking" and devising, Species X came to use more and more of his mental powers, and in time these faculties became more developed. Following the Doctrine of Use and Disuse, Species X shed off bodily parts which he no longer needed for his survival as his mental capabilities expanded and he was able to devise better ways of procuring food. In time, Species X evolved and metamorphosed into Modern Man.

Although Modern Man or Man now is totally different in physical form and mental abilities, it stands to reason that he bears the stamp of his ape forebears. Darwin called this stamp "rudiments." In the *Descent of Man*, Darwin devoted a large portion of Chapter I discussing these rudiments that Man inherited from his ape ancestors. They are the sense of smell, hair, and reproductive system.

With regards to the sense of smell, the lower organic forms such as dogs and horses have acute or "highly developed" sense of smell because these species needed this capability to survive. It was used as a defense mechanism. By being able to readily smell an enemy, the said species could protect themselves by escaping or vanquishing the enemy. But as these species evolved to higher forms and branched out into apes and Man, this capability became less acute because it was no longer needed for survival. Man needed the sense of smell mainly for aesthetic purposes such as inhaling fragrant odors or recalling people and places for emotional reasons. In the *Descent of Man*, Darwin postulated that the sense of smell is one of the rudiments that Man inherited from his distant ancestors:

> In those animals which have this sense highly developed, such as dogs and horses, the recollection of persons and of places is strongly associated with their odor; and we can thus perhaps understand how it is, as Dr. Maudsley has truly remarked that the sense of smell is "singularly effective in recalling vividly the ideas and images of forgotten scenes and places."[32]

Another rudiment that Man inherited from the lower organic forms is hair. In the same book, Darwin discussed this physical characteristic:

> Man differs conspicuously from all the other Primates in being almost naked. But a few short straggling hairs are found over the greater part of the body in the man, and fine down on that of a woman . . . There can be little doubt that the hairs thus scattered over the body are the rudiments of the uniform hairy coat of the lower animals.
>
> I am informed by Sir James Paget that often several members of a family have a few hairs in their eyebrows much longer than the others; so that even this slight pe-

---

[32] Charles Darwin, *Descent of Man*, 18-19.

> culiarity seems to be inherited. These hairs, too, seem to have their representatives; for in the chimpanzee, and in certain species of Macacus there are scattered hairs of considerable length rising from the naked skin above the eyes, and corresponding to our eyebrows; similar long hairs project from the hairy covering of the superciliary ridges in some baboons.[33]

Like the sense of smell, hair as a rudiment inherited from the lower forms is no longer essential for Man's survival. And following the Use and Disuse concept, Man has totally, or almost, shed off this physical characteristic because he no longer needs it to survive. To borrow Darwin's words "Man differs conspicuously from all the other Primates in being almost naked," because, unlike the Primates who need this covering to protect themselves from the cold, Man is not subject to this same element. He has a brain that can devise better ways and means of protecting himself from the cold such as making warm coats that he can easily put on and take off for comfort.

The last rudiment that Darwin discussed is the reproductive system. Here he gave a detailed and lengthy account of how the pattern of the structure of bodily parts of various organic forms including Man is similar--because these parts were formed on the "same ideal plan." Likewise, the embryos of these organic forms, "can at first hardly be distinguished from each other" because they all evolved from a common progenitor. Darwin wrote in the same book:

> The reproductive system offers

[33] Ibid., 19.

various rudimentary structure . . . The homological construction of the whole frame in the members of the same class is intelligible, if we admit their descent from a common progenitor, together with their subsequent adaptation to diversified conditions. On any other view, the similarity of pattern between the hand of a man or monkey, the foot of a horse, the flipper of a seal, the wing of a bat, etc., is utterly inexplicable. It is no scientific explanation to assert that they have all been formed on the same ideal plan. With respect to development, we can clearly understand, on the principle of variation supervening at a rather late embryonic period, and being inherited at a corresponding period, how it is that the embryos of wonderfully different forms should retain, more or less perfectly, the structure of their common progenitor. No other explanation has ever been given of the marvelous fact that the embryos of a man, dog, seal, bat, reptile, etc., can at first hardly be distinguished from each other. In order to understand the existence of rudimentary organs, we have only to suppose that a former progenitor possessed the parts in question in a perfect state, and that under changed habits of life they became greatly reduced, either from simple disuse, or through the natural selection of those individuals which were least encumbered with a superfluous part, aided by the other means previously indicated.[34]

---

[34] Ibid., 24-25.

In conclusion, Darwin explained why Man has these rudiments from the lower forms--he inherited said rudiments from these lower forms who were his ancestors, and Man and these ancestors descended from a common progenitor. Darwin wrote:

> Thus we can understand how it has come to pass that man and all other vertebrate animals have been constructed on the same general model, why they pass through the same early stages of development, and why they retain certain rudiments in common. Consequently we ought frankly to admit their community of descent; to take any other view is to admit that our own structure and that of all the animals around us, is a mere snare laid to entrap our judgment.[35]

In essence, what does this explication of inherited rudiments mean? To Darwin, it means that Man descended from the lower organic forms through linear evolution by natural selection per se. God or the Supreme Being did not create him. Through time, Man evolved from the ape (Homo erectus) without divine or supernatural intervention. With regards to mental powers, Darwin believed that there is no basic difference between the brains of apes and men. It's true that Homo erectus and some of the intermediate species were governed by instinct. But in the course of time, the instincts in Species X (remember him?) developed into intelligence as he struggled against nature, subdued it, and eventually won ascendancy over it. As the brain of Species X became more developed, he shed the rudiments he inherited from his ancestors because he no longer needed these rudiments to sur-

[35] Ibid., 25-26.

vive. Expansion of the brain's capabilities is inversely proportional to the disappearance or shedding of bodily parts no longer needed. So, while the brain of Species X was developing, the body parts no longer needed for survival were slowly shed off. In the process of shedding these rudiments, his physical form changed gradually until he metamorphosed into another species--Homo sapiens. Darwin wrote:

> My object in this chapter (Chapter III) is to show that there is no fundamental difference between man and the higher mammals in their mental faculties . . . A high degree of intelligence is certainly compatible with complex instincts, and although actions, at first learnt voluntarily can soon through habit be performed with the quickness and certainty of a reflex action, yet it is not improbable that there is a certain amount of interference between the development of free intelligence on instinct--which implies some inherited modification of the brain.[36]
>
> The difference in mind between man and the higher animals, great as it is, certainly is one of degree and not of kind.[37]

One other distinguished scientist who criticized linear evolution regarding Man's origin was Robert Foley, a professor at Cambridge University. He wrote:

> The long-held view that humans are unique, totally new type of organism requires a special type of explanation, dif-

---

[36] Ibid., 67, 69.
[37] Ibid., 130.

> ferent from that used in understanding
> the rest of the biological world.[38]

Foley was, of course, referring to the Sumerian clay documents which postulated that extraterrestrial beings from an advanced civilization may have been involved in the emergence of humans because their existence could not be explained by the method used in the understanding of the "biological world." The notion that extraterrestrials were responsible for the creation of humankind had been floating around since the discovery of the Sumerian documents in the early 1800s. Foley, and some fellow scientists, was of the opinion that this subject was worth exploring in the hope that an explanation could be arrived at regarding the existence of Man who he believed was "a totally new type of organism."

Needless to say, the supporters of the ape/man theory dismissed Foley's view as outlandish. Darwin brushed off the extraterrestrial theory of Man's origin, saying:

> It is only our natural prejudice, and arrogance which made our forefathers declare that they were descended from demigods.[39]

He continued to hold this belief until his death in April 1882.

---

[38] Roger Lewin, *Human Evolution*, 4.
[39] Charles Darwin, *Descent of Man*, 26.

Darwin published *Descent of Man* in 1870. To date, the "connecting links" have not been discovered. The Sumerian clay documents explain why these links will never be discovered--because there aren't any. Man was not a product of linear evolution but was created by extraterrestrial beings for purposes of their own.

How did the Sumerian clay documents come to be discovered? What spurred financiers and archaeologists to expend large amounts of money and time to excavate these clay documents?

A lot of theories had been put forward as to what force propelled these individuals to find these documents. One theory was love for money. A lot of money could be made from these artifacts. Museums all over the world were willing to pay a lot of money for these souvenirs from Man's distant past. Another theory or reason was fame. Seekers of fame were quite willing to sacrifice an arm or leg, not to mention fortunes, to see their names in print as discoverers of this and that relic of Man's history. The third reason given as to what brought about the excavation of large volumes of the Sumerian clay documents was Man's search for the truth of his origin--the search for his Creator.

Among the three reasons given, the third held sway as the most compelling reason that explains why financiers and philanthropists financed archaeological expeditions to unearth more Sumerian clay documents. Man wanted to know who he was, where he came from, and where he was headed.

Up until the late eighteenth century, Man believed that the Bible was a valid explanation of reality. He believed that the good Book was a true guide of his existence, his thoughts and actions.When Copernicus found that the earth was merely a planet revolving around the sun and not the center of the universe, thereby diminishing Man's significance in the scheme of things, Man still clung, for the most part, to the Bible as the guide or "map" of his existence, and the repository of Truth. But it could not be denied that the seed of doubt had been planted. As he pondered this doubt in his subconscious, he began to wonder if the Bible indeed contain the Truth or set of truths regarding his existence.

As time passed and as the Industrial Revolution ushered in the age of machines, more knowledge was obtained about these "truths." Fossils were found in search of fuel for these machines. These fossils indicated that they belonged to a group of animals that no longer existed, contradicting the story of Noah about all the pairs of animals brought aboard the ark. The seed of doubt that was planted with Copernicus' heliocentric findings finally took root in Man's mind and made the Bible suspect as the repository of Truth. Its very foundation was shaken. Man doubted if he was indeed the master who had "dominion over all living things on earth" as the Bible stated and he began to question its authority as "the" guide or map of his existence. Its credibility was further weakened when Charles Darwin made a pronouncement that Man descended from the lower animals and was not created by God or the Supreme Being for a lofty purpose.

Needless to say, this pronouncement, along with the discovery of fossils and Copernicus' findings, impacted the general public. People were plunged into a precarious neutral zone--tottering between holding on or letting go of their religious view. The final blow came with the appearance of the Sumerian clay documents. It was the final straw that broke the camel's back. Whatever shred of belief that man clung to as lord and master of God's creation was finally severed as the Sumerian clay documents were discovered.

The Sumerian clay documents were first discovered in the Middle East in the early 1800s. These documents were made of clay, a kind of soil used in pottery, vases, etc. When treated with a certain substance from a tree, the clay becomes sturdy and durable and resistant to corrosion and decay. It was for this reason that the scribes or official historians of Sumer used clay to inscribe their history/culture because these documents were meant to be handed down to posterity. The Sumerian clay documents found verified fossil findings regarding the truth about a segment of Man's history. Fired up by this information, Man was spurred on and was compelled by an inner desire to know more, if not all, the bits and pieces that comprised the Truth of his existence. It was for this reason that vast fortunes were spent to excavate these documents--to find the Truth about Man's existence. They found the Truth all right. But it was not the kind of Truth they expected to find--far from it.

The Sumerian clay documents told man HOW and WHY he was created. Further, these documents resolved a major objection to Darwin's ape/man theory regarding the "connecting links," which he believed, will eventually be discovered. It had been over a hundred years since Darwin made that statement. To date, the missing links or "connecting links" had not been found. The Sumerian documents postulated that these links would never be discovered--because there aren't any. Man was not a product of linear evolution as Darwin stated, but, according to the documents, was created by extraterrestrial beings for purposes of their own.

The land of Sumer, which was located in lower Mesopotamia, was discovered by archaeologists some two hundred years ago. Its culture began and flourished about 6,000 B.C. Unlike ancient Egypt, which perished about 2,500 years ago but whose civilization was never obliterated, the existence of Sumer was totally phased out. Sumerians no longer exist as a people. They were conquered by the Babylonians sometime in 2170 B.C. Three reasons are given as to why knowledge about Sumer and its people were not known until two hundred years ago.

The first reason is attributed to the nature of the land: "in Sumerian times buildings were almost entirely of mud-brick, either sun-dried or kiln-baked."[40] Sumer had no large stones to protect itself from erosion when floods and sandstorms ravaged the land. After it was conquered by the Babylonians, the conquerors deserted the city and left it to the mercy of the natural elements. Soon thereafter, the city ceased to exist. The second reason as to why Sumer was not discovered earlier is that Sumer was a city located in Mesopotamia which was not readily accessible to explorers. Leonard Cottrell wrote about this land's geography in the same book: "A hundred years ago there were only two ways of reaching it [Mesopotamia], on horseback over the dangerous overland route from the east Mediterranean coast, or through Turkey and Armenia; or by the long sea journey round Africa to Basrah on the Persian Gulf. Even if one reached Basrah there was still a journey of many weeks, mainly on horseback before one reached Baghdad and Mosul [or any city of Mesopotamia].[41]

The third reason is that Mesopotamia was vulnerable on all sides. It had no barriers from any of the four directions unlike Egypt which "was protected by well-defined frontiers; the Mediterranean on the north, the deserts on east and west. The only threat was from Nubia in the south, but although the Nubians once conquered Egypt in a time of decadence,

---

[40] Leonard Cottrell, *The Quest For Sumer*, 20.
[41] Ibid.,20.

the Pharaohs usually succeeded in keeping their southern neighbors in check. Mesopotamia, however, enjoys no such protection; the country is open on all sides, and thus had to endure a succession of invasions from the north and east."[42] (See Fig. 2 for location of Sumer)

For several centuries, the cities of Mesopotamia were ravaged by conquerors, namely, the Elamites, Mongols, Arabs, Assyrians, and Babylonians. Sumer, the most notable of these cities because of its rich artifacts and superior culture, was pillaged the most. After its destruction by the Babylonians had been completed and the dust had settled, one significant event occurred. Some of the Babylonian monarchs, who were individuals of vision and appreciated cultural refinement, adapted the Sumerian way of life and made its culture their own. This phenomenon was common in ancient times. A case in point was the conquest of ancient Greece by the ancient Romans. The ancient Greeks had a far superior culture to that of Rome. As a result, when the Romans conquered the Greeks, Rome was swallowed culturally by Greece, that is, the Romans adapted most aspects of Greek life, including the Greek polytheistic religion. So, Greek gods such as Zeus (the king of the gods), Aphrodite (the goddess of beauty), Athena (the goddess of wisdom), Hermes (the messenger of the gods), etc. were given Roman names. Zeus became Jupiter, Aphrodite became Venus, Athena was Minerva, Hermes was Mercury, and so on and so forth.

The records of Sumerian culture are inscribed in clay tablets. Sumerologists or scholars of Sumerian culture refer to these tablets as the "Sumerian clay documents." From these clay tablets or documents, we know how the Sumerians looked like and what kind of society they had. Closer exami-

---

[42] Ibid.

nation of these documents revealed this culture was not like any culture we have known.

Although the land of Sumer was discovered in the 1800s and a few of these tablets were unearthed, it wasn't until the 1900s that archaeologists, financed by wealthy entrepreneurs for reasons mentioned earlier, were able to excavate a substantial amount of these clay documents and make the existence of Sumer known to the public. As more of these clay tablets were discovered and later translated into English and other languages, a totally new world view, depicted by the Sumerians, emerged. Samuel N. Kramer wrote in his book *History Begins at Sumer*:

> One remarkable fact is that only a century ago nothing was known even of the existence of these Sumerians in ancient days . . .
>
> Yet today the Sumerians are one of the best-known peoples of the ancient Near East. We know what they looked like from their own statues and steles scattered throughout several of the more important museums in this country and abroad . . .
>
> Moreover, Sumerian clay tablets by the tens of thousands (literally) inscribed with their business, legal and administrative documents crowd the collections of these museums.
>
> We actually have a large number of Sumerian clay documents on which are inscribed the literary creations revealing Sumerian religion, ethics and philosophy.[43]

---

[43] Samuel N. Kramer, *History Begins At Sumer*, xx-xxi.

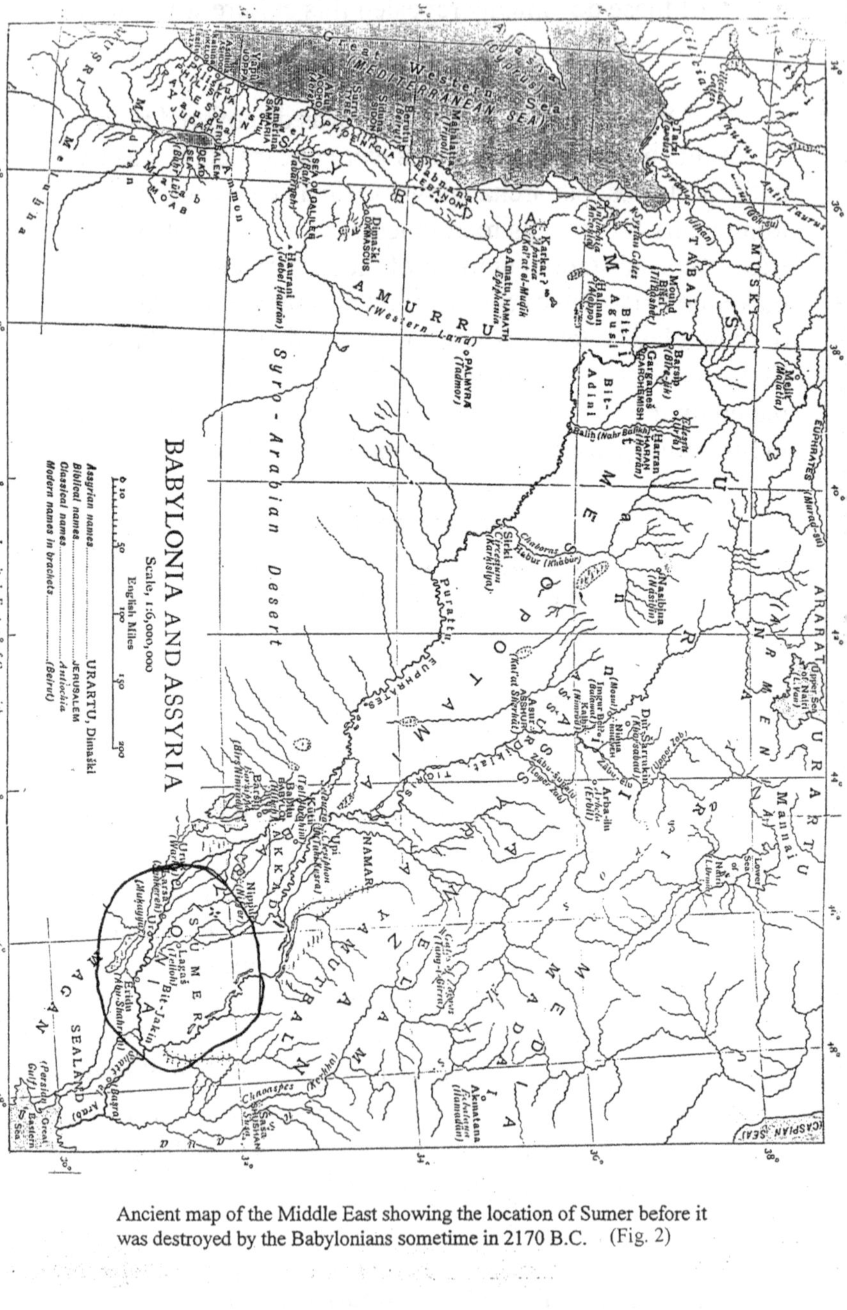

Ancient map of the Middle East showing the location of Sumer before it was destroyed by the Babylonians sometime in 2170 B.C. (Fig. 2)

Perhaps, the most interesting and thought-provoking aspect of Sumerian culture is its religious literature. This literature relates the account of HOW and WHY Homo sapiens or humans were created. It challenged Charles Darwin's theory that "man is descended from some lower form, notwithstanding that connecting links have not hitherto been discovered." The Sumerian clay documents explain why the missing links or "connecting links" will never be discovered--because there aren't any. Homo sapiens or Man was created by extraterrestrial beings or Sumerian gods for purposes of their own.

There are two prevailing versions as to HOW and WHY Man was created by the Sumerian gods. Both versions are in agreement as to WHY and HOW of creation. The Sumerian gods created Man because they needed slaves to do their work; he was created by manipulating Homo erectus' genetic structure through the scientific process called genetic engineering. The only difference in these two versions are the beings who were actually involved in man's creation.

The first version of Man's creation is that of Zecharia Sitchin. In his book, *The 12th Planet*, Sitchin narrated the origin of humankind. According to this book, about 450,000 years ago humanoids from an advanced civilization of Nibiru, the twelfth planet of our solar system, scanned the universe for gold which they needed for survival on their home planet. With the use of their sophisticated instruments, they found what they were looking for on earth. Their rocket ship splashed down somewhere in the marshland of Sumer and there they started the exploration of the precious metal. Their leader was Ea whose name was changed later on to Enki, which meant Ruler of Earth, to reflect his position in this mission. In this version, he is described as a great scientist and builder of pyramids and astronomical devices, and the conversion of marshlands to mining site is attributed to him.

A passage from a Sumerian clay tablet depicted Enki's landing on the marshes of Sumer and his conversion of this marshland into a habitable dwelling:

> Enki, the king of the Abzu, in his majesty, speaks up with authority
> "I am the lord, I am one whose command is unquestioned, I am the foremost in all things
> At my command the stalls have been built, the sheepfolds have been enclosed
> When I approached the earth, there was a high flood
> When I approached its green meadows,
> The heaps and mounds were piled up at My word
> I built my [house] a shrine, in a pure place, I called it with a good name
> My house--its shade stretches over the snake marsh . . .
> The lofty marshland, my favorite spot
> Stretches out its arms to me . . .
> Sing sweet song, cause the river to rejoice."[44]

Enki had a brother named Enlil. Both brothers were in charge of the Earth Project. Enlil supervised the Command Center; Enki the mining operations. (Fig.3)

After a while, mining in the water proved to be unfruitful. It took the worker gods a lot of time to extract the precious metal and the amount extracted was scanty, for Su-

---

[44] Samuel N. Kramer, *The Sumerians*, 3.

Sumerian seal showing the god Enki, Chief Mining Officer, commenting on a gold ingot presented to him by a worker god. In those times a gold ingot had a hole in the middle through which a rod was inserted. (Fig. 3)

mer was mostly a marshland and had practically no minerals and stones. There was only one way to increase production--mine in land, in Africa where gold was abundant. And they did. The base of their gold mining operations was moved from Sumer to Africa. For thousands of years the worker gods labored for the metal which was shipped to their planet.

As time went by the need for gold became more pressing. The atmospheric layer of their home planet was quickly dissipating. Large quantities of gold were needed. To meet this pressing need, the production quota was raised. It proved to be too much for the worker gods. They complained to their leader, Enki, and told him they wouldn't work anymore. It was bad enough toiling underneath the hot, sultry earth for the metal but to double the production was adding insult to injury. A passage from a tablet reproduced in the *The 12th Planet* described the worker gods' distress at having to do the strenuous work:

> When the gods, like men,
> Bore the work and suffered the toil
> The toil of the gods were great,
> The work was heavy,
> The distress was much.[45]

A confrontation between Enki and Enlil ensued. A council of the gods was convened. What would they do? To abandon the mining would mean certain death to the inhabitants of Nibiru. The council voted to continue the mining. The problem was: who would do it? Enki offered a solution. Let the apeman do it. Science tells us that about this time an apelike creature named Homo erectus roamed widely in the continent of Africa and other places. Fossils of this species

---

[45] Zecharia Sitchin, *The 12th Planet*, 331.

have been found in Africa, Java, and Algeria. Since there were no other feasible options, the council agreed to let the apeman replace the worker gods and continue the gold mining.

To enable Homo erectus to perform the task of gold mining, his intelligence needed to be upgraded and his physical make-up altered because Homo erectus was just an ape. He was described as "hairy, shaggy, a roaming nomad who knows neither people nor land, garbed in leaves like one of the green fields, feeding on grass and living among the animals." Accordingly, Ninhursag, the Chief Medical Officer, who was also the sister of Enki and Enlil, performed the task of transforming Homo erectus. She extracted an egg from a female Homo erectus, mixed it with the sperm of a worker god, and implanted the embryo in the womb of a birth goddess. The result of this union was the first Homo sapiens. Modern science confirms this genetic restructuring—Man's body, including his brain was redesigned from that of an ape to understand commands and carry out tasks: (Fig. 4)

> Anatomical comparison suggests strongly that our bodies are apes' bodies redesigned as bipeds . . . Bipedal walking, increased intelligence . . . all contribute to differences between our own and apes' skulls, brains, jaws, and teeth. In relation to body size, man's brain and braincase are far larger than an ape's and the brain is more highly organized, with relatively bigger frontal parietal and temporal lobes, collectively concerned with thinking and social behavior, and producing and understanding speech.[46]

---

[46] David Lambert, *The Field Guide to early Man*, 86.

Sumerian tablet depicting the goddess Ninhursag showing off her creation, the *first* Homo sapiens (Fig. 4)

The gods called him "primitive worker." After the primitive worker passed the test for gold-mining operations, the gods decided to replicate him. He was going to replace the worker gods and mine the gold under the earth. So, several female Homo erectus were subjected to the same procedure and with the assistance of birth goddesses, a workforce of primitive workers was formed. But, as in the case with hybrids, these workers could not procreate because they were sterile, and had short life spans. Ninhursag and her assistants had to repeat the procedure a number of times to replace the primitive workers as they died.

After a while, she felt the process was too time-consuming and tedious. She had to do something. She consulted her scientist brother Enki. After a lot of discussion, weighing the pros and cons, they decided to give the primitive workers the gift of procreation or "knowing." To achieve this goal, Both Enki and Ninhursag had to do a lot of experiments in genetic engineering. Since it was a trial and error process, they made several errors before they were able to produce the perfect worker. One of these errors was Neanderthal man. He is described as "an uncouth creature . . . the hands are large and coarse, and lack the delicate play between thumb and fingers which is found in Homo sapiens. The large brain is singularly defective in the frontal region. It is clear that Neanderthal limbs and brains were incapable of performing those delicately skilled movements that are the distinct prerogative of Homo sapiens." Aside from the Neanderthals, fossils of this failed experiment at creating the ideal worker were found in various parts of the world. They are: Petralona Man from Greece: Arago Man from southwestern France; Steinheim Man from Germany; Broken Hill Man from Zambia, and so on. Despite many differences among these individual specimens, they all have two things in common: they are more advanced that Homo erectus, having larger brains, and more primitive than Homo sapiens. Because of the varying anatomy of these specimens, "anthropologists labeled these fossils collectively as archaic sapiens."[47] Other specimens of this failed attempts are found in mythology: the satyr, a creature with the upper body of a man and the torso of a horse; the sphinx, who had the face of a man, the body of a lion and the wings of an eagle; and others. All these creatures were not figments of our human ancestors' imagination. They were real flesh and blood specimens of a failed experiment. Two accounts from the Babylonian texts, *Enuma Elish: The Seven Tablets of Creation* and *Babylonian Legends of Creation*, that are adaptations from the Sumerian texts, describe these specimens:

> From my land I went not forth, and I
> did not give them battle
> A people who had the bodies of birds,
> of the hollow men who had the
> faces of ravens
> Had the great gods created
> And in the ground the gods created a
> dwelling place for them.[48]

> There appeared men, some of whom were furnished with two wings, others with four, and with two faces. They had one body but two heads, the one that of a man, the other of a woman; and likewise in their several organs both male and female. Other human figures were to be seen with the legs and horns of goats; some had horses' feet; while others united the hindquarters of a horse

[47] Richard Leakey, *The Origin of Humankind*, 83.
[48] L.W. King, *Enuma Elish: The Seven Tablets of Creation*, 143.

> with the body of a man, resembling in shape the hippocentaurs. Bulls likewise were bred there with the heads of men; and dogs with fourfold bodies, terminated in their extremities with the tails of fishes; horses also with the heads of dogs.
>
> There were creatures in which were combined the limbs of every species of animals. In addition to these, fishes, reptiles, serpents and other monstrous animals, which assumed each other's shape and countenance, all of which were preserved *delineations* in the temple of Belus [Enlil] at Babylon.[49]

After several more attempts, Enki and Ninhursag finally came up with the right combination. This was accomplished by removing an egg from a female primitive worker, uniting it with Enki's sperm, and placing the fertilized egg in the womb of a birth goddess. The result of this procedure was the first modern Homo sapiens. They called him Adapa or Adam. Enki and Ninhursag repeated the same procedure, but this time they manipulated the process in such a way that the offspring would be female. They called her Titi or Eve. Both Adapa and Titi or Adam and Eve were placed in the garden of the gods to tend it. Conducting more experiments on the hybrids, Enki finally discovered a way for his two children to procreate. In the Bible, the humans' ability to reproduce is symbolized by Adam and Eve's eating the fruit of the Tree Of Knowledge of Good and Evil.

The second version of the Sumerian creation is that of several translators/writers. The foremost of these writers

---

[49] *Babylonian Legends of the Creation*, British Museum, 1921.

is Herman Wohlstein. In his book *The Sky God An-Anu,* Wohlstein credits Anu, Enlil, Enki and Ninhursag for Man's creation. He writes:

> An, Enlil, Enki and Ninhursag had
> The black headed (mankind) created.[50]

Whereas in the first version, only Enki and Ninhursag are involved in Man's creation, in the second version, Anu and Enlil also play active roles in the event. Both versions are in agreement as to why Man is created: to do the worker gods' task. A line from the *Babylonian Legends of the Creation* states: "He [Enki/Ea] fashioned mankind for the service of the gods, and he set the gods free."

The story goes as follows: An/Anu as head of the Council of Gods, elected his son Enlil to appoint the god most suitable to the task of creating the creature who would replace the worker gods. After a conference between Enlil and the other great gods, they decided that Enki is the best candidate for the job. In this version of the creation story, Enki is depicted as a great administrator/manager who knows the skills of each of the gods and goddesses. He appoints Ninhursag, the Chief Medical Officer, to create the human worker. A passage from the *Encyclopedia Britannica* describes Ninhursag as:

> Goddess of birth, she is not only the
> goddess
> Of animal birth but the Mother of all
> children,
> A mother goddess figure.[51]

---

[50] Herman Wohlstein, *The Sky God An-Anu*, 4.
[51] "Ninhursag," *2001 Encyclopedia Britannica Inc.*

Not only was Enki a great manager, but he was a lover of mankind as well. Findel and Geller commented of this trait of Enki in their book *Sumerian Gods and Their Representations:*

> There is another well-known trait of Enki's character which may be relevant in this context, viz, his never ceasing love and care for mankind. For what Enki is actually doing here is laying the conditions for the good life on earth, which is taken to be identical to the idealized Sumerian way of public life. These conditions consist of a well-ordered organization . . . The ideal organization molded by Enki out of the world as it already was, is foremost beneficient to Man.[52]

In the second version, Enlil was depicted as the most powerful god, next to his father An or Anu. Enlil had the power of life and death over the human creatures. He was the same god who decreed that Man should perish in the Flood. The following is a description of Enlil's power:

> Enlil, whose command is far-reaching,
> whose word is holy
> The lord whose pronouncement is unchangeable who forever decrees destinies
> Whose lifted eye scans the lands,
> Whose lifted beam searches the heart

[52] I.L. Findel and M.J. Geller, *Sumerian Gods and Their Representations*, 130.

> of all the lands,
> Enlil who sits broadly on the white dais,
> on the lofty dais
> Who perfects the decrees of power, lordship and princeship
> The earth-gods bow down in fear before him
> The heaven-gods humble themselves before him.[53]

Compared to the Sumerian version regarding Man's creation, the Biblical version pales beside it. Whereas the Sumerian account is clear and convincing as to WHY and HOW Man was created, in the Bible, both issues are vague and ambiguous.

According to the Book of Genesis, Man (Adam) was created to be the lord and master of all living things on earth. A passage illustrating this, reads:

> And God said, "Let us make man in our image, after our likeness, and let him have dominion over the fish of the sea, and over the fowl of the air, and over the thing that creepeth upon the earth."[54]

Why would God or the Almighty want a *human*, who is a finite and limited being, be the lord and master of all living things on earth? What was His reason? If the WHY of the Creation story is perplexing, the HOW is equally confusing because the images are in symbols and the reader is left

---

[53] I.L. Findel and M.J. Geller, *Sumerian Gods and Their Representations*, 130.

[54] "Genesis", 1: 2 (King James Version)

to interpret the meaning for herself/himself. An example is the following: Man was made from dust and he became alive only when the Lord breathed into his nostrils—"From dust thou art to dust thou shalt return," the famous line says. The images of "dust" from which Man was made and the "breathed into his nostrils" to give him life are symbols which no two readers would interpret the same way. The result, of course, is confusion.

Further, Man and Woman were not created as equals. According to the Book of Genesis, Adam was created first. Eve was made only as a helpmate or companion to Adam when God saw that Adam was lonely. The following is the passage that illustrates this:

> And the Lord God said, "It is not good that the man should be alone; I will make him a helpmate for him."
>
> And the Lord God caused a deep sleep to fall upon Adam and he slept; and He took one of his ribs and closed up the flesh instead thereof.
>
> And the rib, which the Lord God had taken from man, made he a woman, and brought her unto the man.[55]

Taken literally, the above passage meant that Woman is inferior to Man because she is only an ancillary or additional being. She does not have a separate existence of her own. She was created not as an equal to Man but as a helpmate or vassal to him.

Because both the WHY and HOW of the Biblical ac-

---

[55] Ibid., 2: 3.

count of the Creation is shrouded in symbolism and figurative language, God's purpose in creating Homo sapiens is apt to be misinterpreted. If symbols are hard to interpret, figurative language is difficult to understand because it requires an excellent knowledge of a language and an expertise in literary devices.

Strange to say, among the parallel events in human history found in the Sumerian documents and the Bible such as Man's creation, the ability to procreate, the Flood, and the second coming of the gods or Redeemer, only the Creation tale of the Sumerian text is changed *blatantly* in the Bible. In the Bible, not only is an inferior position assigned to Woman but her being is further debased by imputing to Eve that she disobeyed God by eating the forbidden fruit from the Tree of Knowledge of Good and Evil and persuading Adam to eat also. As punishment for her disobedience and her persuading Adam to sin, she is depicted as the cause of Man's or humanity's loss of Paradise or the good life, and plunging humanity or Man into the hard life by having to eat from the sweat of his brow. This Biblical account is a *deliberate* change from the Sumerian documents. Nowhere in the Sumerian text does it say that Titi is inferior to Adapa or that she was the cause of the humans' hardship suffered as a result of a god's wrath.

If this *change* did not catch the attention of most readers it is because this was done in the systematic use of symbols and figurative language. As we mentioned earlier, no two readers will interpret a symbol or a figurative phrase or sentence the same way. Why did the Biblical scholars systematically use symbols and figurative language in the Creation narrative?

Several hypotheses are given but perhaps the most logical, if not compelling reason, is because there wasn't enough data to work on. The Biblical writers had to use symbols and literary devices because they didn't have enough factual information in hand. As in literature, a writer who doesn't have sufficient material about a subject would emphasize style or form rather than content. Style includes use of symbolism and sophisticated literary devices such as points of view and flashback, and figurative language such as simile, metaphors, onomatopoeia, alliteration, and so on. A book whose format is manner of presentation or style would use the foregoing devices. On the other hand, a book whose emphasis is on content would use straightforward language. The difference between the two books is that the first book is an exercise in aesthetics; the second, a direct commentary of life.

A substantial amount of the Sumerian clay documents have been discovered. Some of these documents narrate the history of Man's creation up to the second coming of Man's creators, the Sumerian gods. These gods will make a final assessment of humanity, and decide whether humanity will be allowed to continue to live or to perish. In the Bible, this event is called Armageddon.

It's not surprising that several significant events in human history, such as Man's creation, his ability to procreate, the Flood, and the second coming of a redeemer (God or god) have parallels in the Bible and the Sumerian clay documents—*the books of the Bible are merely excerpts of the Sumerian clay documents*. Samuel Kramer, a noted Sumerologist, comments on the source of the Bible and other great books of the world such as the Greek Iliad and Odyssey, the Hindu Rig-Veda, and the Iranian Avesta:

> . . . the Bible, which contains the literary creations of the Hebrews, the Iliad and Odyssey, which are filled with the epic and mythic lore of the Greeks, the Rig-

> Veda which contains the literary products of ancient India, and the Avesta which contains those of Iran--none of these literary collections were written down in their present form before the first half of the first millennium B.C. Sumerian literature, inscribed on tablets dating from approximately 2000 B.C. antedates these literatures more than a millennium. Moreover, there is another vital difference. *The texts of the Bible, of the Iliad and Odyssey, and of the Rig-Veda, and Avesta, as we have them, have been modified, edited, and redacted by compilers and redactors with varied motives and diverse points of view* (italics mine). Not so our Sumerian literature; it has come down to us as actually inscribed by the ancient scribes of four thousand years ago unmodified, and modified by later compilers and commentators.[56]

Except for the Creation narrative that was modified deliberately and systematically in the Bible, all other events in the Bible, in essence, are faithful renditions of the Sumerian texts.

As time passed, following the humans' creation and grant of the ability to procreate, the humans proliferated and outnumbered the gods. It appeared that the humans, having been bestowed the ability to reproduce, had become sex fiends. They overindulged in their carnal appetite and did too

---

[56] Samuel Kramer, *Sumerian Mythology*, 20.

many evil deeds. They regressed into their animal origins. Because of this regression into the ape or animal state, the humans had made their bodies inhabitable to the divine spirit, which was contained in Enki's sperm. The god's essence or spirit would not protect Man because his body was derived from an animal. The Book of Genesis similarly states: "And the Lord said, 'My spirit shall not always strive with man, for that he is flesh'."[57] To the minds of the gods, humanity had rendered themselves undesirable specimens and had to be destroyed, bodily, to release the spirit. The Deity (Enlil) in the Sumerian text says:

> And the Deity saw that the wickedness
> of Man was great on the earth
> And that every desire of his heart's
> thoughts was only evil every day
> And the Deity repented that He had made
> Man upon the earth and His heart
> grieved
> And the Deity said: "I will destroy the
> Earthling whom I have created
> off the face of the earth."[58]

The Bible says the same thing. In the Book of Genesis, it states:

> And God saw that the wickedness of man was great in the earth, and that every imagination of the thoughts of his heart was only evil continually.
>
> And it repenteth the Lord that He had made man on the earth, and it

[57] "Genesis", 6: 7.

[58] Zecharia Sitchin, *The 12th Planet*, 376.

> grieved Him at His heart.
>
> And the Lord said, "I will destroy man whom I have created from the face of the earth; both man and beast, and the creeping things, and the fowls of the air; for it repenteth me that I have made them."[59]

But God did not totally destroy humanity. Both the Bible and the Sumerian text agree that the Cosmic God or the Supreme Being allowed humanity to "continue to exist." The Book of Genesis says:

> But Noah found grace in the eyes of the Lord . . .
>
> And said unto Noah, "the end of all flesh is come before me; for the earth is filled with violence through them; and behold I will destroy them with the earth.
>
> Make thee an ark of gopher wood . . . and of every thing of all flesh two of every sort shalt thou bring into the ark to keep alive with thee, shall be male and female."[60]

In the Sumerian text, Enki is instructed by the Creator of All to save humanity through his (Enki) son Ziusudra. In a dream vision the messenger of the Creator relayed to Enki the message from the Creator:

> In the dream-vision there appeared the

---

[59] "Genesis", 6: 5-7.

[60] Ibid., 8: 19.

image of a man, bright and shining like the heavens;
And as the man approached, Enki saw that the white-haired Galzu he was!
In his right hand an engraver's stylus he was holding
And in his left hand a tablet of lapis lazuli, shining smooth, he held
And as he approached near enough by Enki's bed to stand, Galzu spoke up and said;
Unwarranted your accusations against Enlil were, for only the truth he spoke;
And the decision that as Enlil's Decision will be known, not he but Destiny decreed
Now into your hands Fate take, for the earthlings the Earth will inherit;
Summon your son Ziusudra, without breaking the oath to him the coming calamity reveal
A boat that the watery avalanche can withstand, a submersible one, to build him tell,
The likes of which on this tablet to you I am showing;
Let him in it save himself and his kinfolk,
And the seed of all that is useful, be it plant or animal also take
That is the will of the Creator of All.[61]

[61] Zecharia Sitchin, *The Lost Book of Enki*, 219-20.

The reason behind the Creator's decision to allow Man to continue to exist was practical. Enki, the benefactor of mankind, persuaded his brother Enlil that humans were needed to do the work of the gods. Aside from mining gold, the gods needed the humans to do the task of providing the gods' needs such as clothing, shelter and food. We see this issue illustrated several times both in the Bible and in the Sumerian texts.

In the Sumerian tablet, "The Dispute between Cattle and Grain," the lesser goddesses Lahar and Ashman were created to provide clothing and food for the greater gods, but the two goddesses failed in their assigned task. Since they were ineffectual, the task was therefore given to the humans. A passage from the tablet read:

> In those days, in the creation chamber
> of the gods
> In their house Duku, Lahar and Ashman
> were fashioned
> The produce of Lahar and Ashman,
> The Anunnaki [gods] of the Duku
> eat but remain unsated;
> In their pure sheepfolds shum-milk,
> the gods
> The Anunnaki of the Duku drink but
> remain unsated;
> For the sake of their pure sheepfolds,
> the gods
> Man was given breath.[62]

In the Bible, the Lord was prevailed upon to let Man continue to exist upon beholding Noah's offering. After the

[62] Samuel Kramer, *The Sumerians*, 221.

water receded and land was visible once more, Noah got out of the ark and slaughtered and roasted the best fowl and animals in his livestock and offered them to the Lord. Savoring the aromatic smell of Noah's "offerings," the Lord was reminded that the gods do need the humans to provide the gods' basic necessities, among other things, and there and then decided to let Man live. I said gods because, like the Sumerian text, the Bible specifically stated that other gods, aside from the Lord/God, created Man. The Book of Genesis stated:

> And God said, "Let *us* make man in our image after our likeness . . ."[63]

Whereas in Man's creation other gods were involved, in determining Man's fate (whether he should live or die), the Lord acted *alone.* He decided to let man continue to live. The passage that illustrates this is the following:

> And Noah builded an altar unto the Lord; and took of every clean beast; and of every clean fowl and offered burnt offerings on the altar.
>
> And the Lord smelled a sweet savour; and the Lord said in his heart, "I will not again curse the ground any more for man's sake; for the imagination of man's heart is evil from his youth; neither will I again spite any more every thing living, as I have done."[64]

The same is true in the Sumerian documents. Enlil

---

[63] "Genesis", 1: 2.
[64] Ibid., 8: 10.

acted *alone* when he decided to let Man continue to exist. His brother, Enki, persuaded him that humans were needed to provide food and other necessities of the gods. In the epic *Atra Hasis: The Babylonian Story of the Flood*, (an adaptation of the Sumerian text) Enlil learned about the thirst and hunger of the gods during the Flood. He was told that on board their spaceship the gods witnessed the Flood that ravaged the land for many days while weeping and suffering from hunger and thirst:

> The Anunnaki, great gods were sitting in thirst, in hunger
> Ninti wept and spent her emotion; she wept and eased her feelings.
> The gods wept with her for the land.
> She was overcome with grief; she thirsted for beer.
> Where she sat, the gods sat weeping; crouching like sheep at a trough
> Their lips were feverish of thirst, they were suffering cramp from hunger.[65]

Upon learning of the gods' plight, it dawned upon Enlil that Man and the food he produced was essential to the gods' survival on earth. Thereupon, Enlil decided to heed Enki's advice and make friends with the survivors of the Flood. He took Atra Hasis to the home of the gods: "The logic of the situation and Enki's persuasion prevailed; Enlil made his peace with the remnants of Mankind and took Atra Hasis in his craft up to the Eternal Abode of the Gods."[66]

---

[65] Zecharia Sitchin, *The 12th Planet*, 398.
[66] Ibid., 400.

The Flood which decimated nearly the entire human race brought about a radical change in the gods' handling of the humans.

Enki, the creator and father of humankind, angrily lashed at his brother Enlil for letting almost all of humanity perish in the Flood. Were it not for his (Enki's) ingenuity, all of humankind would have been obliterated from the face of the earth. Enlil responded saying that he had no hand in the catastrophe. Man, himself, was responsible for the tragedy--his wickedness had doomed him. Back and forth the brothers bickered with no resolution in sight. This animosity dragged on for a while.

Meantime, the human remnants of the Flood reproduced, multiplied, and replenished the earth. As the human population grew, it became mandatory to establish some form of government, similar to antediluvian times, to maintain peace and order. But it was not business as usual. The old infrastructure was gone--it perished with the Flood. The old system was a dictatorship where Enlil had absolute power. He had the power of life and death over the humans. Although *by law*, he was in charge of the Command Center and not of human affairs, in *fact*, he was the ruler of humanity. This same situation existed in ancient Rome during the time of Julius Caesar. We might even say that the humans copied this system of government from their gods. Anyway, two thousand years ago, the Roman Empire, *by law*, was governed by the Senate but in *fact* was ruled by the military. So, when the military commander, Julius Caesar, returned from his conquest of Gaul, Rome's last enemy or the last to resist Rome's power, the senators conspired and killed him. They might just as well have done so because, given Caesar's popularity with the Roman masses and his insatiable thirst for power, he probably would have abolished the Senate and proclaimed Rome a military state with himself as the Supreme Commander.

After the near demise of humanity, Enki opposed his brother and challenged his leadership. The bitter rivalry between these two powerful siblings erupted into an open and violent conflict. A meeting of the council of gods was convened. Its purpose was to set up guidelines on how humanity was to be governed and to officially proclaim An/Anu's heir. Enlil's performance as ruler of humanity was closely scrutinized and evaluated, and he was found wanting. His father, An/Anu, the god in charge of the gods, recommended to the Council that his son Enki, and not his other son Enlil, should succeed to the throne. The recommendation was approved and this decision of An/Anu naturally created a civil war because, by the Nibiruan Law of Succession, Enlil was the rightful heir of An/Anu since he was the offspring of An/-Anu by his sister, while Enki's mother was no relative of An/Anu at all. Nibiruan Law of Succession states that the child born from the ruler by his blood kin has a greater claim to the throne than the child who's parent is no blood kin of the ruler at all. Enlil, therefore, was legally the rightful heir to the throne. An/Anu, with the support of the Council, rejected Enlil as the successor to the Nibiruan throne and Ruler of Earth because he was not moral. He had committed too many immoral and reprehensible acts. Some of these acts were his rape of the goddess Ninlil, his unilateral decision to destroy several cities, and his plan to decimate humankind. In contrast, his brother Enki was by far better in character. He was a suitable ruler of gods and humans. As manager, he had proven his expertise when he appointed the right god or goddess to help develop Sumer and other cities as home for gods and humans. Likewise, he was a suitable ruler of the humans because, for one thing, he was the creator/father of Adapa and Titi (Adam and Eve.) He, naturally, would look after the well-being of his children's descendants--the human race. The Council voted Enki as An/Anu's successor and

Ruler of Humanity. Enlil was reduced in rank and was to take orders from Enki. Needless to say, Enlil was enraged and blatantly defied the council. Civil war erupted.

Perhaps, this was the most recorded war in human history. To be sure, there were battles and wars among the gods before--for one reason or another. But these conflicts mainly concerned the affairs of the gods--their struggle for power, etc. But this time, the war concerned mainly the fate of humans. Enlil believed that humanity should have all perished in the Flood. To his mind, Man was an undersirable specimen because he was made from a beast. In view of this, Man would regress into the ape or animal state like he did before and pollute planet earth with his evil and wicked ways. Enlil thought that humanity, therefore, should be exterminated. But the council of gods thought otherwise. The Creator of All concurred with the Council. True that Man was made from the beast or ape but he bore within him the spirit or essence of the great god Enki. Adapa and Titi were issues of Enki's sperm and the egg of a female primitive worker. A female primitive worker was an offspring of the ape, Homo erectus, and a worker god. Since Man carried within him Enki's essence, humans were spiritual beings. As spiritual beings they had the right to co-exist among other spiritual beings unless the Council and Creator of All declared otherwise.

The Council believed that when the Sumerian gods returned to earth, they would make a final assessment of the human condition. The Council would then present its findings to the Creator of All who would render a final judgment whether humanity should be allowed to continue to live or die. Until then, humanity should be allowed to exist, the Council ruled.

Enlil rejected Enki's appointment as An/Anu's heir

and Ruler of Humanity. In consequence, he and his followers were driven out from the abode of the gods.

Enlil's rebellion and fall is recorded in almost all literatures of the world. In Persian literature, he is Ahriman, the evil god and ruler of the demons. In Islam, he is known as Iblis, the leader of the fallen angels. Perhaps, the most vivid account of Enlil's defiance and expulsion from the world of the gods is recounted in the Jewish and Christian religious literature. Both religions depict Enlil as Lucifer, the archangel, second to God in command. In both versions, Lucifer was cast out from heaven because of his sin--intellectual pride. He rejected Man because he was a lowly creature made from dirt. As punishment for his sin, Lucifer was cast into hell along with his followers. John Wickersham in his book *Myths and Legends of the World*, wrote:

> Jewish and Christian traditions offer similar explanations for the devil's origin . . . Satan was originally an archangel, one of God's most divine or blessed creations. His name is given sometimes as Samuel but more often as Lucifer.
>
> Some accounts say that God cast the archangel out of heaven because he would not honor Adam . . . When the jealous archangel refused to acknowledge "a lowly thing made of dirt" God punished his pride by throwing him down into hell. There, as Satan, the fallen archangel ruled over a kingdom of devils, former angels who had followed him in his fall.[67]

[67] John Wickersham, *Myth s and Legends of the World*, 23.

Enki's overall plan was to train humans how to govern themselves. When the humans could effectively govern themselves the gods would withdraw from the scene and humans would be given full rein to run their lives. To achieve this objective, Enki set about training humans in various occupations. After finishing her/his training, the human was placed in a position he/she was suited for.

When the gods deemed that humanity was ready to govern themselves, they slowly withdrew from human affairs and transferred the reins of power to human administrators to run the cities but the gods still, essentially, owned the land. The administrative head of the city was a High Priest selected by the god or goddess himself/herself. The god's official residence was the temple where the administration of a city was performed by the city functionaries. The following is a description of this setup:

> The principal god or goddess of the city was its ruler . . . the god owned the land, the High Priest administered the city on his behalf--like the manager of an estate owned by an absentee landlord . . . Sumerian society was cooperative, with every detail carefully planned. In Sumerian documents the workers on the temple estates, priests, herdsmen, fishermen, gardeners, etc. were referred to as "the people of the god x."[68]

As time passed and the human population grew, a new form of government was established. Administration of a city was transferred to a governing body called the "assem-

---

[68] Leonard Cottrell, *The Quest For Sumer*, 160.

bly." The tasks of the High Priest and other temple functionaries were exclusively devoted to "religious" duties. In most cases, they had none or little say on the affairs of government.

The assembly form of government, which was democratic in substance, was later changed to monarchial. It happened this way: in the democratic form of government, all citizens were free. The governing body or assembly consisted of an upper house of elder statesmen and a lower house of men. The assembly was headed by an "ensi." He presided over the two houses. Situations or issues important to the city were voted upon by the two houses. The military merely served to preserve peace and order. This was the normal course of affairs in the cities.

As time went by, life became more complex. With increased population there was more need for land and resources. As a consequence, a struggle for these resources and land developed and became violent. If these internal problems were not enough, the cities were attacked by barbaric tribes from neighboring lands.

It should be mentioned here that these barbaric nomads were people who left the cities when the gods had abdicated their thrones and transferred the reins of power to the humans. The majority of these humans stayed in the city, and although they were free to run their lives, they were still under the guidance and protection of their god. Some humans chose to leave the city to be on their own. Being on their own, they were not under the jurisdiction of a god or goddess. Without the watchful eye of a deity, these humans regressed into the animal or barbaric state, reminiscent of antediluvian days. History calls these humans "barbarians." These barbarians are characterized as unruly and undisciplined people. They had no system of government or culture of their own. They simply adapted the systems of cities they attacked and conquered. Some of these rootless hordes were the Huns, Visigoths, and Vandals. In his book *The Barbarians*, Tim Newark describes these savage people. He wrote:

> Barbarian was a derogatory name, an insult. The Greeks used the word for all those people living beyond their frontiers . . . Later, the Romans and their heirs, applied the word to European and Asian races living outside their empires who did not inhabit sophisticated urban settlements. These Barbarians were viewed as wild, savage people . . . Barbarians are seen as the antithesis of civilization. Destroyers. For this reason, it has been said, our advanced societies must always be on their guard against Barbarian elements or lapses into Barbarism.[69]

As internal problems mounted and as attacks from barbarians became more frequent, the democratic form of government ceased to be viable. Both houses of the assembly were impotent to resolve these internal problems and the barbarian attacks which threatened the very life of the city itself. There was pressing need for a strong leadership. Seeing that the democratic form of government was no longer working, the god of the city changed it. The governing body or assembly was dissolved and a one-man rule was instituted. The head of this new form of government was the king. In his book *The Sumerians*, Noah Kramer describes this transition from the democratic to the monarchial. He wrote:

---

[69] Tim Newark, *The Barbarians*, 7.

> In early days political power lay in the hands of free citizens and a city governor known as "ensi" . . . In case of decisions vital to the city as a whole, these free citizens met in a bicameral assembly consisting of an upper house of elders and a lower house of men. As the struggle between city states grew more violent and bitter, and as the pressures from the barbaric peoples to the east and west of Sumer increased, military leadership became a pressing need and the king . . . came to hold a superior place.[70]

The king was personally selected by his god. Although the king had broad powers, encompassing both secular and religious domains, he did not rule by his volition but by the will of his god. He took orders from his god and was accountable to his god alone. The king's task, primarily, was to administer justice to his people--this was the ideal. But human nature being what it was, and still is, the human king was all too susceptible to the human vices of greed, avarice, covetousness, oversize ego, etc. Abuses were committed by men of power and, in most cases, by tacit approval of the king. Since the god was not always around to monitor the king's doings, these abuses remain unchecked for some time. Kramer wrote:

> Artisans and apprentices were reduced to abject poverty and had to beg for their food. Blind men--presumably prisoners of war and slaves who had been blinded in order to prevent them from at-

[70] Noah Kramer, *The Sumerians*, 74.

> attempting to escape--were seized and put to watering the fields like animals and were given only enough food to keep them alive. The rich--and the supervisors, were getting richer and richer at the expense of the less fortunate citizens by forcing them to sell their donkeys and houses at low prices and against their will. The indigent, the poor, the orphaned and the widowed were mistreated and deprived in one way or another of what little they had by men of power and influence.[71]

Upon his return to the city, the god normally ordered the king to appear before him and report on activities during his absence. After this meeting between god and king, the god conferred with his intelligence network which was comprised of trusted personnel from different departments of the kingdom. The abuses were investigated. If the abuses were confirmed, the god took action by replacing the incumbent corrupt king with a presumed just one. Like his predecessor, the new king made an agreement or covenant with the god that he would administer justice to his people and would see to it that no powerful man would commit an unjust act toward his fellow citizens. If the new king was just, he normally was allowed to rule for a long time by his god. Upon his death, his eldest offspring succeeded to the throne, and the line of succession flowed from father to eldest child for as long as the kingdom prevailed or as decreed otherwise by the god. To date, some societies, like England, still practice this form of government whereby the monarch believes that his or her power emanates from his/her god, and that he/she is

---

[71] Ibid., 81.

above the law. This doctrine is called the Divine Right of Kings. In his book entitled *The Divine Right of Kings*, John Neville Figgis wrote:

> An Englishman will see in the Divine justification for the English law of succession . . . From the consideration of the popular acceptance and practical object of the doctrine, some obvious conclusions may be drawn. First, it seems clear that so general and enthusiastic a faith must have been the expression of deep-seated instincts; secondly, that a doctrine so fully elaborated and yet so eminently the product of a definite epoch must have been the result of a long chain of historical causes, and that it must have been formed to meet real needs.
>
> Kings are accountable to God alone. Monarchy is pure, the sovereignty being entirely vested in the king, whose power is incapable of legal limitation. All law is a mere concession of his will and all constitutional forms and assemblies exist entirely at his pleasure.[72]

> We will still believe and maintain that our Kings derive not their title from the people but from God; that to Him only they are accountable; that it belongs not to subjects, either to create or censure, but to honor and obey their sovereign, who comes to be so by a fundamental hereditary right of

[72] John Neville Figgis, *The Divine Right of Kings*, 5-6.

> succession, which no religion, no law, no fault or forfeiture can alter or diminish.[73]

While the "kingship" form of government was taking place on earth, major changes were happening in the world of the gods. The great gods or Council of Gods was needed back at their home planet. Matters of grave importance necessited their return. Much as they wanted to stay to oversee their human creatures and check the stewardship of the king, they had no choice but to leave. Before their departure, the Council convened a meeting to formulate a system of laws or codes for the king on how he should justly rule his subjects. Some of these codes or laws are *The Laws of Ur-Nammu, Lipit-Ishtar's Code,* and *The Code of Hammurabi.*

Perhaps, *The Code of Hammurabi* is the most well-known among these laws probably because a substantial amount of tablets have been excavated and were found to be organized to be called a system of laws or code. This code consists of an Introduction or Prologue, the body or text of 282 laws, and an Epilogue.

In the Prologue, King Hammurabi makes an announcement that the god of justice, Shamash, gave him these laws:

> . . . to cause righteousness to prevail in the land, to destroy the wicked and the evil, to prevent the strong from plundering the weak, to go forth like the sun over the black-headed race [humans], to enlighten the land and to

---

[73] Ibid., 73.

> further the welfare of the people.[74]

The text, consisting of 282 consecutively numbered laws, list offenses and their corresponding penalty. The following are examples of law violation pertaining to justice and negligence, and their corresponding punishment:

Justice:
1) If a man has accused a man and has cast against him an accusation of murder and has not proved it against him, his accuser shall be put to death.

Negligence:
55) If a man has opened his ditch for irrigation and has been slack and has consequently caused the water to carry away his neighbor's field, he shall pay corn correspondingly to the crop of the field adjoining it.[75]

Although some of these laws denote the idea of an "eye for an eye" kind of justice, the general import or spirit of these laws is to promote legal and social justice. In the Epilogue, King Hammurabi addresses his successors to implement his code of laws to their subjects or suffer a curse he will bear upon them if they don't. In essence, the Epilogue is similar to the Prologue because both state that the purpose of the code is to exhort the ruler to administer justice:

> . . . that he may thus make straight the way for his black-headed ones [humans], that he may judge their causes and decide their decisions, that he may pluck out the evil and the wicked from his land, and

---

[74] J.M. Pouvis Smith, "The Code of Hammurabi", trans. by D.D. Luckenbill, *The Origin and History of Hebrew Law*, 181.
[75] H.W. Saggs, *The Greatness that was Babylon*, 209.

make the flesh of his people glad.[76]

The gods thought that these codes would enable humankind to establish a fair and just society. They believed that these guidelines would foster civilizations of law and order.

Countless millennia have passed since the Cosmic God or Creator of All decided to allow humanity to continue to live. Thousands of years have passed since the gods left with a promise to return. When the gods return to earth the Council of Gods will evaluate the humans in terms of their adherence to the righteous path. This is the final judgment where the Council of Gods, presided by the Creator of All, will rule whether humanity should perish or be allowed to continue to exist. In the Bible, this event is called Armageddon whereby Jesus Christ will cast the wicked ones to hell and take the righteous with Him to heaven.

[76] Ibid., 206.

How reliable is the Sumerian text in its account of Man's creation, his development and eventually his fate? Is there scientific evidence to support the veracity of all these issues? The answer is yes. We have scientific data to support all of the above.

To begin with Man's creation--scientific data strongly suggests that Man was not a product of linear evolution as Charles Darwin postulated but was created by extraterrestrial beings (Sumerian gods) by genetically altering Homo erectus' body, including his brain, for purposes of their own.

Linear evolution is defined as the uninterrupted development of life forms or species without divine or supernatural intervention. All life forms, that is, plants and animals including Man, was a product of linear evolution through the process of natural selection and not created by God or the Supreme Being. Nature selects what species are to perish and which species are to survive and form other distinct species. In his book *Descent of Man*, Darwin stated unequivocally that Man, like the lower organic forms, was a product of linear evolution, that he "descended from some lower form notwithstanding that connecting links have not hitherto been discovered." He wrote:

> Even if it be granted that the difference between man and his nearest allies is as great in corporeal structure as some naturalists maintain, and although we must grant that the difference between them is immense in mental power; yet the facts given declare, in the plainest manner, that man is descended from some lower form, notwithstanding that connecting links have not

hitherto been discovered.[77]

The "connecting links" issue was the hottest topic in the linear evolution controversy. It was so because this issue was the reason that this controversy could not be resolved. These "connecting links" are the intermediate species that separate the ape Homo erectus and Homo sapiens. Furthermore, these missing links spell the difference in mental capabilities between Man and his ape relative. Man has evolved greatly in intelligence as evinced in his abilities to build civilizations and to travel to the moon and other planets. His ape progenitor is still an ape and is still governed mainly by instinct. Darwin believed that the "connecting links" would show how instinct in the ape developed into intelligence in Man.

Darwin published *Descent of Man* over a hundred years ago. To date, the "connecting links" have not been discovered. Since they have not been found, science cannot establish as fact or confirm Darwin's theory that Man evolved from the ape or that he (Man) is a product of linear evolution.

Since the publication of *Descent of Man* and up until the discovery of a substantial amount of the Sumerian clay documents in the 1900s, the public hung in a limbo regarding Man's origin. Was Man created by God or is he a descendant of apes? The world took sides. One side was for the former, the other side, for the latter. As mentioned earlier, this controversy was the famous Creationist versus Evolutionist debate. The Creationists were believers of the view that the human race was created by God or the Supreme Being; the Evolutionists, on the other hand, were partakers of the theory

---

[77] Charles Darwin, *Descent of Man*, 24.

that humanity came into existence through the process of linear evolution per se.

For more than a century the battle between Creationists and Evolutionists has raged on. A breakthrough occurred in the 1900s with the appearance of a substantial amount of the Sumerian clay documents which narrated Man's creation, his development, and eventual fate. These documents explain that the "connecting links" will never be discovered because there aren't any. Man is not a product of linear evolution but was created by extraterrestrial beings as slave labor. The mental capabilities of Man's brain and the time frame that he appeared on earth following Homo erectus invalidate the "connecting links" theory. There are no missing links or "connecting links" because Man did not evolve from Homo erectus. Man is a modified Homo erectus, genetically engineered from the laboratory of extraterrestrial beings or gods.

In Darwin's time and up to the discovery of the Sumerian clay documents, Man's brain was the core of the "connecting links" theory regarding the linear evolution controversy. Richard Owen, the most eminent comparative anatomist in Darwin's generation, initiated the brain/hippocampus issue in 1861 in response to Thomas Huxley's statement "that no significant difference between ape and human brains exist." The brain/hippocampus debate of 1861 followed the Oxford Debate of 1860 between the Bishop of Oxford and Thomas Huxley. Whereas in the Oxford Debate of 1860, Richard Owen remained in the background and used the Bishop of Oxford as his mouthpiece to oppose Darwin's ape/man theory, in the brain/hippocampus debate of 1861, Owen came out into the open and confronted Huxley head on. He refuted Huxley's statement and averred that a significant difference between ape brain and human does exist because the ape brain lacks the *hippocampus* in the human brain. This controversy captured the attention of the public and before the month was over, most periodicals in Britain picked up the brain/hippocampus issue. A satire of the Owen- Huxley debate appeared in a British periodical called *Punch* and was reprinted by Cybil Bibby in his book entitled *Scientist Extraordinary: The Life and Scientific Work of Thomas Huxley 1825-1895*. It read:

Then Huxley and Owen
With rivalry glowing
With pen and ink rush to the scratch;
'Tis Brain versus Brain
Till one of them's slain;
By Jove! It will be a good match!

Next Huxley replies
That Owen lies
And garbles his Latin quotation
That his facts are not new
His mistakes not a few,
Detrimental to his reputation

To twice slay the slain,
By dint of the brain
(Thus Huxley concludes his review)
Is but labour in vain
Unproductive of gain
And so I shall bid you "Adieu"[78]

Little was known scientifically about the function of the brain in the 1860s, especially about the *hippocampus*, so the debate did not create a lot of ripples outside of the British Isles. Now that more is known about the brain and the *hippo-*

---

[78] Cyril Bibby, *Scientist Extraordinary: The Life and Scientific Work of Thomas Huxley*, 47

campus as a result of experiments and clinical studies, the *hippocampus* issue has been resurrected. (Fig. 5)

To begin with, Webster defines brain as "the portion of the vertebrate central nervous system that constitutes the organ of thought and neural coordination, includes all the higher nervous centers receiving stimuli from the sense organs and interpreting and correlating them to formulate the motor impulses, is made up of neurons and supporting and nutritive structures, is enclosed within the skull, and is continuous with the spinal cord."[79]

The brain, both ape and human, has two hemispheres, the left and right hemispheres. Each hemisphere consists of four major parts, namely: (Fig. 6)

1) frontal lobe – in charge of movement and speech; destruction of this lobe results in paralysis

2) parietal lobe – the "feeling" part; receives impulses from the sensory organs – injury to this lobe may cause the person not to feel, say, the exact spot of a bee sting; the person will still feel the sting but won't know whether she was stung in her arm or leg – only an intact parietal lobe will tell the person where exactly the sting occurred

3) occipital lobe – visual center; injury to this part results in blindness

---

[79] *Webster's New Collegiate Dictionary*, 132.

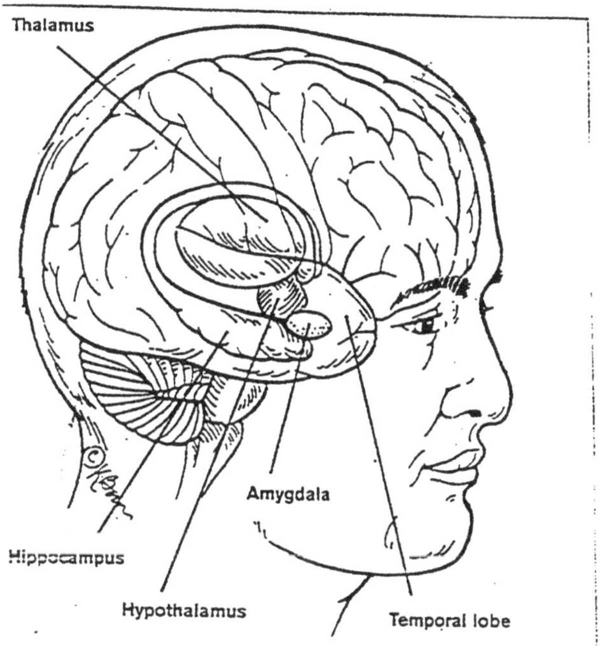

Illustration of the human brain showing the parts of the frontal and temporal lobes. The *hippocampus* which is in charge of storing memories of past events is absent in the ape brain. (Fig. 5)

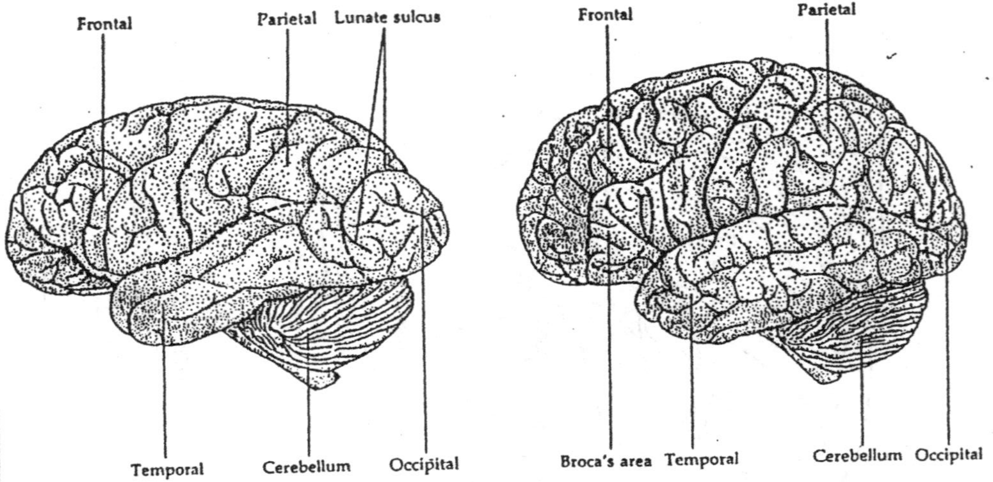

Diagram comparing the ape brain (left) and the human brain (right). *Broca's area* which is associated with speech ability is fully developed in humans but hardly developed in the apes to be noticeable, even through a microscope. (Fig. 6)

4) temporal lobe – associated with hearing and memory; the *hippocampus* which is in charge of storing memories is located in this area

Contrary to Huxley's opinion that ape and human brains do not differ significantly, modern science has established the fact that the two brains do differ significantly because of the *hippocampus*, among other things. The ape's temporal lobe does not have a *hippocampus*, the human temporal lobe does. The *hippocampus* is extremely important because this organ differentiates instinct from reason and ape from Man. The *hippocampus* is an organ in the brain associated with memory and is responsible for storing sense perceptions and events of the past. In his book entitled *Matter and Memory*, Henri Bergson describes the nature and function of memory:

> It [memory] stores up the past by the necessity of its own nature. By this memory is made possible the intelligent, or rather intellectual, recognition of perception already experienced; in it we take refuge every time that in the search for a particular image, we remount the slope of our past. But every perception is prolonged into a nascent action; and while the images are taking their place and order in this memory, the movements which continue them modify the organism and create in the body new dispositions toward action. Thus is gradually formed an experience of an entirely different order, which accumulates within the body, a series of mechanisms wound up and ready, with reactions to external stimuli ever more numerous and more

> varied and answers ready prepared to an ever growing number of possible solicitations.[80]

Because humans have a *hippocampus* which stores past thoughts and experiences, the individual's action is voluntary--voluntary in the sense that he/she can recall memories at will and respond to a given stimulus by choice. In the ape or lower animals, choice of action is not possible because they don't have a *hippocampus*. Action in the lower organic forms is not from choice but instinct.

Without a memory, an organism responds to a given stimulus by instinct, not by choice. This is the fundamental difference between Man and his ape ancestor Homo erectus. Because Man has a memory, his consciousness is capable of making several choices based on his past experiences. Man responds to a given stimulus in various ways by his power of choice. In the lower animals such as ape, reaction to a stimulus is purely instinctive, for animal consciousness is totally different from human consciousness. In his book *Creative Evolution*, Henri Bergson differentiates these two types of consciousness:

> Radical, also, is the difference between animal Consciousness, even the most intelligent, and human Consciousness. For consciousness corresponds exactly to the living being's power of choice; it is co-extensive with the fringe of possible action that surrounds the real action. Consciousness is synonymous with invention and with freedom. Now, in the animal, invention

[80] Henri Bergson, *Matter and Memory*, 264.

> is never anything but a variation on the theme of routine. Shut up in the habits of the species, it succeeds, no doubt, in enlarging them by its individual initiative; but it escapes automatism only for an instant, for just the time to create a a new automatism. The gates of its prison close as soon as they are opened; by pulling at its chain it succeeds only in stretching it. With man, consciousness breaks the chain. In man, and in man alone, it sets itself free.[81]

Likewise, Albert Einstein underscores the importance of memory saying that it is a crucial organ which distinguishes Man from the lower forms such as ants and bees. Memory makes it possible for Man to accomplish works of art, science, engineering and other fields of human endeavor that comprise culture or civilization. Also, memory spells the difference between instinct and choice. In his autobiography *Out Of My Later Years*, Einstein differentiates between instinct and conscious thinking or choice:

> While the whole life process of ants and bees is fixed down to the smallest detail by rigid, hereditary instincts, the social pattern and interrelationships of human beings are very variable and susceptible to change. Memory, the capacity to make new combinations, the gift of oral communication have made possible developments among human beings which are not dictated by bio-

---

[81] Henri Bergson, *Creative Evolution*, 263-64.

> logical necessities. Such developments manifest themselves in traditions, institutions and organizations, in literature; in scientific and engineering accomplishments; in works of art. This explains how it happens that, in a certain sense, man can influence his life through his own conduct, and that in this process conscious thinking and wanting can play a part.[82]

Aside from the *hippocampus*, one other fundamental difference between ape brain and human brain is Broca's area. Richard Owen did not mention this in his debate with Thomas Huxley because during his time the function of the different parts of the brain could not be determined. Now it can. Just as modern science has determined the function of the *hippocampus*, it has also ascertained the role of Broca's area. Broca's area is located in the fontal lobe of the brain and is the seat of understanding and language production. Closer study of the ape brain revealed that Broca's area has been found in the earlier Homo species up to Homo erectus, but, since this organ was not fully developed, the Homo species up to Homo erectus could not produce the universal vowel and consonant sounds in Homo sapiens' speech. Fossils indicate that Broca's area was fully developed in Homo sapiens about 300,000 years ago, confirming the Sumerian clay documents' account that archaic Homo sapiens (modified primitive workers, remember them?) were created by extraterrestrial beings some 300,000 years ago. In his book *Human Evolution*, Lewin, a scientific journalist at Peabody Museum in Harvard University, wrote:

> Only with the origin of archaic

[82] Albert Einstein, *Out Of My Later Years*, 126.

> Homo sapiens, some 300,000 years ago does the fully modern pattern [vocal tract] appear, indicating at least the mechanical potential for the full range of sounds produced by people today.[83]

One may ask, "But didn't Darwin say that it takes a long time for organs to develop and form another distinct species?"

Yes, Darwin said that. But he also said that it takes thousands of generations or millions of years for a species to diverge and form another species as exemplified in the evolution from Australopithecus to Homo habilis then to Homo erectus.

Science considers Australopithecus as the first hominid ape from which Homo erectus, Man's immediate ancestor, evolved. Australopithecus appeared more than five million years ago. He is considered the first hominid in the sense that he was humanlike--he was bipedal or walked on two legs. Phillip Tobias, in his book, *Hominid Evolution*, commented about the Australopithecines as the line from which humans evolved:

> The curious anatomy of the Australopithecines which indicates a curious evolutionary mosaicism as it evolved towards the human condition, is generally held.[84]

Australopithecus had a brain size of about 450 cubic

[83] Roger Lewin, *Human Evolution*, 196.
[84] Phillip Tobias, *Hominid Evolution*, 102.

centimeters. Fossils of this species found in three sites, namely, Taungs, Sterkfontein and Makapansgat in Africa indicated that Australopithecus had a bone-bludgeon culture. This ape was mainly carnivorous and used bones to slaughter its prey. Raymond Dart, in his book titled *The Osteodontokeratic Culture of Australopithecus Prometheus*, wrote about this species method of butchering its prey:

> This uniformity at the three sites [Taungs, Sterkfontein, Makapansgat] caused me to claim the systematic employment by the Australopithecinae of a skilled, predatory, bone-bludgeoned technique.[85]

The hominid species from Australopithecus to Homo erectus had two things in common: bipedalism and, as they evolved, larger brain size.

From Australopithecus, the hominid line continued and branched out to Homo habilis. Homo habilis had a larger brain than Australopithecus. Its brain size was about 600 cubic centimeters--an increase of 150 cubic centimeters from its predecessor. Homo habilis had a less projecting face, shorter arms relative to its legs, and smaller hip bones to allow bipedal walking. He stood about five feet tall with rounded head, had brow ridges, and protruding jaws. Artifacts found along with fossils suggest that Homo habilis used simple stone tools to kill its prey, gather plants for food, and hunt game. This species lived about 3,000,000 years ago followed by the more evolved Homo erectus, Man's immediate ancestor.

---

[85] Raymond Dart, *The Osteodontokeratic Culture of Australopithecus Prometheus*, 2.

Homo erectus branched out from Homo habilis about 730,000 years ago. Homo erectus had a brain size of about 800 cubic centimeters, 200 cubic centimeters more than its predecessor, Homo habilis. Like Homo habilis, the skull of Homo erectus had the same old fashioned features. Fossils found of Homo erectus in Africa, Europe, Eat Asia, and Java strongly suggest that this species had a "shelving forehead, thick brow ridges, a flatter face than ours, big projecting jaws, teeth more massive than our own chin . . . their improved technology including standard toolkits; big game hunting put erectus far ahead of former hominids."[86]

Homo sapiens came into existence about 300,000 years ago, immediately following Homo erectus. Homo sapiens had a brain size of about 1200 cubic centimeters, 400 cubic centimeters larger than Homo erectus. Below is an illustration of the time gaps among Australopithecus, Homo habilis, followed by the more evolved Homo erectus and the appearance of the first Homo sapiens:

1) Australopithecus – appeared about 5 million years ago

2) Homo habilis – came about 3 million years ago

3) Homo erectus – evolved from Homo habilis and came into existence 730, years ago

4) Homo sapiens – the first Homo sapiens appeared some 300,000 years ago

---

[86] David Lambert, *The Field Guide to Early Man*, 116.

The gap between Australopithecus and Homo habilis is about two million years. The gap between Homo habilis and Homo erectus is approximately 2,300,000 years. It took over two million years for Homo erectus to evolve from Homo habilis and two million years for Homo habilis to branch out from Australopithecus and the difference between these three species is their use of tools. Australopithecus, who was mainly a carnivorous ape, used bones to bludgeon or slaughter its prey; Homo habilis used simple stone tools to hunt and gather food; and Homo erectus used a standard toolkit for big-game hunting. All three hunting techniques are merely results of usage of the animal brain which is instinctual or non-thinking organ. All three species are still apes. The gap between Homo erectus and Homo sapiens is only 430,000 years and the difference between these two is *neural*, that is, the capabilities of their brains.

The difference between Homo erectus and Homo sapiens in terms of mental capabilities is tremendous and does not serve the theory of linear evolution. Given the above dates, it's not possible that Homo sapiens was a product of linear evolution because Darwin himself said that it would take thousands of generations or a million years to develop an organ and form a new and distinct species. If it took more than two million years to develop a non-thinking/non-rational brain as in the transition from Homo habilis to Homo erectus, it certainly would take more than three million years to develop a sophisticated brain part such as the *hippocampus* in Man. Yet, in a mere 430,000 years, Homo sapiens branched out from Homo erectus with a brain capable of harnessing the power of the atom and building spaceships to explore the moon and other planets. The gap of 430,000 years between Homo erectus and Homo sapiens means that Homo sapiens is not a product of linear evolution as Darwin postulated in his books, *Origin of Species* and *Descent of Man*.

Richard Leakey, a famous paleontologist, supports this view. He wrote in his book, *The Making of Mankind*:

> Scientists see the large neurological differences between apes and men as indicating that there cannot have been a direct transition, that is, the branching out of Homo erectus to Homo sapiens.[87]

One of the greatest achievements of science in the twentieth century was the emergence of molecular anthropology and molecular biology which gave birth to the serum protein test and Mitochondrial DNA test, respectively.

If the reader recalls, I mentioned earlier in this treatise that Charles Darwin published his book *Descent of Man* in 1870 wherein he unequivocally stated that Man descended from the apes. Needless to say, this statement shocked the world because most people believed in the Genesis story regarding Man's creation. Some thought that Darwin's evolution theory would die a natural death because the Church would find a way to quash this heresy like it did during the Inquisition whereby people who professed such blasphemous statements were burned at the stake. But not this time. The Church was unsuccessful. The ape/Man theory could not be stamped out. The "enlightened" people of the world refused to buckle down and fought religious and secular authorities for their right to explore and investigate conservative and revolutionary ideas alike.

As time passed, the ape/Man theory gained momentum and turned into a global controversy. This controversy

---

[87] Richard Leakey, *The Making of Mankind*, 139.

could not be resolved, decisively, because there was no scientific proof or hard evidence to support either side. Now, there is.

Sometime in the early 1800s, controversial documents, the Sumerian clay documents, were found in the Middle East. Some of these documents confirmed fossil findings regarding the fallacy of the Noah story in the Bible. This information, along with Copernicus' heliocentric finding and Darwin's ape/Man theory, made people question the reliability and authority of the Bible as the "map" or guide of Man's existence. People wanted to know if there were more questionable truths in the Bible. Spurred by the compelling need to know more about these truths in general, and Man's origin in particular, archaeological expeditions were mounted to unearth more of the Sumerian clay documents. The search for more of these documents dragged on for several decades.

The breakthrough occurred in the 1900s with the discovery of a substantial amount of the Sumerian documents. These documents narrated an account of Man's creation, his development, and eventual fate.

As time passed, science was able to come up with ways and means to find out the veracity of the Sumerian documents. We might even say that fueled by the compelling need to know the truth, scientific truth, the scientists looked for a way to resolve the issue of Man's origin, once and for all. Their findings confirmed the claims in the Sumerian clay documents regarding Man's origin, development, and eventually his fate.

Our scientists were able to come up with instruments that would measure and test the various states or condition of the human body in a given time, and its relation to other forms of life. The X-Ray machine can tell the status of a person's lungs, if they're healthy or diseased; the electrocardiogram can tell whether a person's heartbeat is regular or iregular; and the protein serum test can ascertain the relatedness of a human with other organic forms.

In the wake of the Creationist versus Evolutionist controversy, the issue being which explained Man's origin correctly, that is, whether he was created by the Supreme Being or was a product of linear evolution, some scientists thought that they could scientifically resolve the controversy by using the protein serum test on apes and humans to determine their relationship, if there's any.

Protein is a giant molecule composed of a long chain of smaller molecules called amino acids and is the principal substance in the blood plasma of vertebrates. Scientists can construct a phylogenetic tree of, say, vertebrates and depict the evolutionary order of these vertebrates as to *when* each species branched out to other species by the sequencing of the amino acids of the protein of these species. In the serum protein test that was used to determine the relatedness between Primates (a higher order of vertebrates) and humans, scientists used the chimpanzee, gorilla, and ape. Test results of this study revealed that the protein sequencing of the ape differed less with humans than the chimpanzee and gorilla, which means that the ape, presumably Homo erectus, is Man's immediate ancestor and not the chimpanzee or gorilla. Charles Darwin is right in postulating that the ape is Man's immediate predecessor and not any other Primate. In his book, *Descent of Man*, Darwin theorized that apes are closer or more akin to humans than other primates because of their anatomical structure. This theory was confirmed by the serum protein test. The more, if not the most, conclusive of these tests was the one performed by Morris Goodman, a biochemist from Wayne State University. Roger Lewin, a journalist/historian, summed up Goodman's results, saying:

> With Goodman's results on the comparison of serum proteins [of apes and humans] which gets much closer to the fundamental material of inheritance, the science of molecular anthropology was effectively born and the Darwinian view reinstated. Anthropologists had long argued about the identity of the immediate ancestor of humans, and from the genetic evidence it had now become clear. It was some kind of ape--not a chimpanzee but something from which the chimp and gorilla also descended.[88]

Darwin was right in stating that Homo sapiens' immediate ancestor was an ape (Homo erectus), but he was wrong in saying that Man evolved from Homo erectus and that the "connecting links" would prove this.

What Darwin did not know was the fact established by modern science that Man couldn't have branched out from Homo erectus because of the "large neurological differences between apes and Men." The gap of a mere 430,000 years that it took Homo sapiens or Man to branch out from Homo erectus nullifies his linear evolution theory.

Another scientific proof that invalidates the theory of linear evolution and supports the extraterrestrial theory regarding Man's origin is the Mitochondrial DNA test.

DNA tests are used in courts of law worldwide to de-

---

[88] Roger Lewin, *Scientific American Library*, 17.

termine the veracity or truth of evidence. Usually, in a criminal case, the verdict is based on the results of DNA tests because of the infallibility or reliability of these tests.

Deoxyribonucleic acid, better known for its acronym DNA, is the molecule that carries the genes of a person. In common parlance, the DNA molecule is often called the "unit of heredity" because it is the molecule which transmits the physical characteristics of the parent to its offspring.

DNA serves two functions, namely, the general and the particular. In particular, the DNA molecule is the ID of a person because no two people have the same DNA. The DNA of a person is unique--it is his or hers alone. It is for this reason that a murderer can be identified if she/he leaves a part of his/her body, for example, a strand of hair, on the victim's body because that single hair contains his/her DNA. In general, the DNA molecule is the carrier of the species. Through the DNA molecule, scientists can determine the descendants of an individual through the ages. Because while it identifies the individual, the DNA molecule, that is, the mitochondrial DNA, also identifies who this person's descendants are.

The DNA cells in our body contain the nuclear genome and the mitochondrial genome. Unlike the nuclear genome which contains the genes of both parents, *the mitochondrial genome contains only the genes of the mother*. The Mitochondrial DNA test was the first test to determine the mother of modern humans. This test was first conducted sometime in the late 1980s by Douglas Wallace, a scientist from Emory University, and later on by Allan Wilson, also a scientist from the University of California. The mitochondrial DNA of the mother is transmitted from generations to generations, ad infinitum. It was for this reason that Wallace and Wilson used the Mitochondrial DNA test to find out *when* the female progenitor of modern humans came into existence. The test result strongly suggests that the female from whom all modern humans descended lived about 140,000 to 280,000 years ago. In consequence, scientists coined the phrase "Mitochondrial Eve Hypothesis" to give credence to the Biblical account that Eve or Titi (in Sumerian) was the mother of all modern humans. Adam and Eve or Adapa and Titi were created within this time frame. Roger Lewin, in his book *Human Evolution*, wrote about this study. He remarked:

> Mitochondrial DNA is useful for tracking relatively evolutionary events for two reasons: first, the DNA, which codes for 37 genes, accumulates mutations on average 10 times faster than occurs in nuclear DNA; second, unlike an individual's nuclear genome, which consists of a combination of genes from both parents, *the mitochondrial genome comes only from the mother* (italics mine). Because of this maternal mode of inheritance, no recombination of maternal and paternal genes occurs; such a mixture may sometimes blur the history of the genome as read be geneticists. Potentially, therefore, mitochondrial DNA offers a powerful way of inferring population history, unhindered by the genetic fog of recombination.
>
> One of the first significant observations to emerge from this work was that the amount of variation of mitochondrial DNA types in the modern human population throughout the world is surprisingly low--just one tenth of that known among chimpanzees, for instance. One explanation is

> that modern humans evolved very recently, a view that Wallace and Wilson independently supported. A calculation based on the rate of accumulation of mutations of mitochondrial DNA gave a time of origin [of the first viable human female] of about 140,000 to 280,000 years ago.[89]

In the early part of this treatise, I mentioned that extraterrestrial beings or gods landed on earth 450,000 years ago in search of gold that was badly needed by their home planet. It was a matter of life and death that they obtain this metal. When the worker gods refused to continue to mine this metal under the surface of the earth, it became mandatory that replacement workers be found to resume the mining. After finding several candidates for a suitable replacement worker, the gods zeroed in on Homo erectus. The gods believed he was the best candidate for the job of gold mining. The great god Enki and the Chief Medical Oficer Ninhursag altered Homo erectus' genetic structure to enable him to perform the task of gold mining. The result of this procedure was the first Homo sapiens. The gods called him "primitive worker." He was not a viable specimen, that is, he could not reproduce because he was a hybrid and therefore sterile; and he also had a short life span. The goddess Ninhursag and her assistants had to repeat the creation procedure countless of times to replace the short-lived primitive workers. After a while, she felt that the process was too time-consuming and laborious. She consulted her scientist brother Enki, and both agreed that the primitive workers should be genetically restructured to enable them to reproduce. After several attempts at creating the ideal worker, Ninhursag and Enki fi-

---

[89] Roger Lewin, *Human Evolution*, 176.

nally succeeded in creating the first viable humans: Adapa and Titi.

The Mitochondrial DNA test, which indicated that the first viable female from whom all modern humans descended lived roughly about 250,000 years ago, conforms to the Sumerian clay documents' account that Ninhursag and Enki were able to create the first viable humans, Titi and Adapa, at this time period.

Likewise, the serum protein test, which inferred that the ape is Man's immediate ancestor, confirmed Charles Darwin's theory that Man descended from the ape Homo erectus, as ascertained by modern science based on this species fossil record. But contrary to Darwin's linear evolution theory that Man evolved from Homo erectus--*Man is a genetically engineered Homo erectus, created and designed by Enki and Ninhursag to serve the gods.*

It's mind-boggling to think that we were created by a bunch of extraterrestrial beings as slave labor and not by the Supreme Being for a lofty purpose as we were taught in Bible school. Be that as it may, this situation or human condition is not bleak as it appears to be. On the contrary, it is a blessing in disguise that we know the truth about our existence because this truth has set us free. Now that we know who we are and where we came from, we will know what course of action to take to realize our full potential.

In the Sumerian text, the gods promised to return to earth sometime in this century and will render a final judgment whether humanity should perish or be allowed to continue to live. Granted that this promise is fulfilled, three questions are paramount in people's minds: first, will the gods give us a second chance?; second, if we are given a second chance, are we going to continue to live in our present form? and third, what will our fate be?

To answer the first question, let me give my little speech on the "orderliness of the universe." The orderliness of the universe, according to Albert Einstein, is due to the Universal Law which states that all things, animate and inanimate, follow a set of rules established by the Creator/God from the beginning of time. These laws are logical and comprehensible to the intellect. If some of these laws are incomprehensible to humans, it's not because these laws are illogical, but because the human intellect has not yet developed to the point of comprehending these laws.

If the Universal Law is logical, it follows that God or the Creator's judgment will be logically sound. If the Creator deems that the humans' usefulness has been outweighed by their behavior and that they have become a menace and threat to the universal community, the Creator may put an end to humanity's existence. Several learned individuals think along this line. The more eloquent of these writers is Ruth Montgomery who echoes Charles Darwin's natural selection theory. She wrote:

> Man, like the dinosaurs, will be withdrawn if he becomes a menace, rather than merely a pest as of now. Experimentation in creation is still continuing, for nothing is static.
>
> We are always undergoing experimentation. Many early species which became a scourge to others have since disappeared from Planet earth. Man will one day disappear in physical form if he contiues to pollute the earth. Live and let live is the universal law, and when one species assumes a destructive role, his time on earth is short-lived, for unless we are helping others there is no place in God's plan for that species or individual.[90]

It's possible that humans will one day cease to exist. This is not mere speculation. It is a distinct possibility for several species have come and gone. Fossil records prove this. We, too, will pass into oblivion if we prove to be a destructive species and a menace to the universal community. As Montgomery succinctly put it, unless our species overcomes its violent or destructive instinct, "there is no place in God's plan for that species or individual."

But, if we do survive the final judgment and the Creator gives us a second chance, are we going to continue to live in our present form? No, chances are our physical

---

[90] Ruth Montgomery, *The World Before*, 64-65.

form will change as our mental capabilities continue to increase.

Change is the permanent condition of life. As the great naturalist Jean Baptiste Lamarcke put it, this change is due to the inner force within every organic form of life, including Man, which drives it forward or backward as it constantly interacts with its environment. In humans, social psychology calls this inner force the drive for self-actualization.

Social psychologists tell us that there is in each of us an authentic self which makes us unique from everyone else. The manifestations of this authentic self defines the true nature of our being and tells us who we really are. Abraham Maslow, an eminent practitioner in this field, describes the authentic self thus:

> We have, each one of us, an essential inner nature which is instinct-toid, intrinsic, given, natural . . . I include in this essential inner nature instinctual basic needs, capacities, talents, anatomical equipment, physiological or temperamental balances, prenatal and natal injuries. This inner core shows itself as natural inclinations, propensities or inner bent. Whether defense and coping mechanisms, style of life, and other characterological traits are all shaped in the first few years of life . . . This raw material very quickly starts growing into a self as it meets the world outside and begins to have interaction with it.[91]

---

[91] Abraham Maslow, *Toward A Psychology of Being*, 212.

Throughout a person's life there is a constant struggle for dominance among the three selves of an individual, namely, the social self, the fantasy self, and the authentic self. The social self is who you want people to think you are; the fantasy self is who you imagine you are; the authentic self is who you really are. Unless one lets the authentic self take command of one's life, a person will never find inner peace, and worse, that person might experience psychological problems. Maslow wrote in the same book, *Toward A Psychology Of Being*:

> If this essential core [authentic self] of the person is frustrated, denied or suppressed, sickness results, sometimes in obvious forms, sometimes in subtle and devious forms, sometimes immediately, sometimes later.
>
> It seems quite clear that personality problems are loud protests against the crushing of one's psychological bones, of one's true inner nature. By going through life without resolving this problem the individual is likely to pay years later in neurotic and psychosomatic symptoms of various kinds.
>
> General illness of the personality is seen as any falling short of growth or of self-actualization. And the main source of illness, although not the only one, is seen as frustrations of the basic needs, of expression of the [authentic] self.[92]

The force that propels the authentic self for domin-

[92] Ibid., 214-15.

ance is what Maslow calls the drive for self-actualization. Maslow further states that "in the hierarchy of human values," self-actualization is Man's greatest need or primary directive. Being Man's primary directive, the authentic self will constantly struggle for supremacy against the social and fantasy selves until it has achieved its goal. Once the authentic self has achieved its objective, then and only then will the individual find inner peace and happiness. However, in the struggle for supremacy among the three selves, a person may stumble and destroy himself/herself psychologically and end up on a psychiatrist's couch. If the patient tries hard enough to recover his/her balance, he/she may become whole again, but not for long. Unless the authentic self is in control, the vicious cycle repeats itself again and again through the end of a person's life.

A Roman senator once said that the law of life is: either you grow or die. How the individual responds to everyday challenges or stimuli will determine how she/he will fare in life. Everyday, the individual is bombarded by different stimuli from his/her environment--from home, family, work, and people and situations outside of home and work. Some of these stimuli can be very stressful. In the course of time, individuals who cannot cope with these stresses will get sick, deteriorate, and eventually die. This is a fact. Modern medicine has established the fact that heart and other organ ailments are caused by an individual's failure to cope with the stresses in his life. Those who survive and live long lives will make an important contribution to knowledge. Scientists will study the lives of these individuals and determine how such individuals are able to triumph over the challenges or obstacles in their lives. Their findings will enable scientists to find ways and means on how to make the human body better-equipped to tackle the stresses from its environment or problems in life and lengthen its life span.

With the passage of time, as Man acquires more knowledge about himself and his world, his mental capabilities will expand and, logically, his brain will grow and increase in size. To accommodate a larger brain, Man's head will, naturally, become bigger. This is a view commonly held by scientists--that Man's physical form will change as his mental powers increase. He will still look human (or humanoid, if you will), but his head will be larger in proportion to his body, for (following the Use and Disuse Doctrine), he will have less and less use of his body as his mental capabilities continue to increase.

In response to the third question, what, eventually, will our fate be, the answer is: we will be gods like the Sumerian gods who created us--*if* we survive the final judgment.

This is not fantasy or wishful thinking. It is a distinct possibility, if not a probability. In point of fact, this is the fate of all spiritual beings. All spiritual beings are potential gods. This is the very reason why Enlil wanted to eliminate the human race. He was afraid that humans, being Enki's descendants, would replace him and his heirs in the chain of being. Let me explain.

To begin with, there is nothing awesome or overwhelming about the Sumerian gods. Their "supernatural" powers stem from their knowledge of the components and forces of the universe and their ability to control and manipulate these components and forces. How are they able to accomplish this feat? Through time. Eons and eons of time. These gods are highly evolved beings. This means that their species is hundreds of thousands or probably millions of years old. In this long life span, they have been able to study and figure out the laws that govern the forces of the universe such as gravity, electromagnetism, etc., and the relationship among its components, namely, matter, space, and time. From knowing how these forces and components operate, these beings have been able to develop a highly advanced technological civilization, because hand in hand with the knowledge of the workings of the universe is, of course, expertise in the various sciences such as astronomy, physics, genetic engineering, and so on and so forth.

The being who can control and manipulate the universe's forces and its components has the following capabilities or abilities: the ability to fly (because he has conquered gravity), and the ability to alter his molecular structure--meaning he can assume any form he wants, or be visible and invisible at will. This ability is derived from his knowledge of the laws governing matter and energy and how they can be changed from one form to the other. Matter is congealed energy and, in its liquid and solid forms, is visible; matter when released from its congealed form is energy and is invisible. By knowing the nature of matter and energy, along with the ability to alter one's molecular structure, anyone would have the power to be, say, an eagle or a lion, or become invisible at will. Likewise, an expertise in the science of genetic engineering would enable one to create life forms--just as the Sumerian gods created us, Homo sapiens.

Capabilities or abilities to fly, to become visible or invisible, to alter one's molecular structure, to create other forms of life, etc., are nothing but application of the laws of nature. It's not magic. It's science.

Granted that our species survives for a long time, say, hundreds of thousands of years, we will be able to evolve and develop to our full potential. In time, we will be like the Sumerian gods and do the "supernatural" things that they did. Is this possible? Yes, it is. The following is an explanation of how the human species can become like the Sumerian gods, that is, again, *if* the human race survives the final judgment.

As I mentioned earlier, the universe is orderly because it is governed by the Universal Law which states that all things, animate or inanimate, follow a set of rules established by the Creator from the beginning of time and these laws are logical and comprehensible to the human intellect. The following is what Einstein said about the universe or the world:

> The most incomprehensible thing about the world is that it is comprehensible.[93]

The universe is comprehensible because it is governed by the Universal Law and this Law, which is a set of the laws of nature, can be understood because it is based on facts, and as such, is reducible to a mathematical equation. To understand a law of nature, the scientist either uses Newton's or Einstein's method.

In his *Laws of Motion*, that is, planetary motion, Newton arrived at these laws by formulating a mathematical equation based on Kepler's observations about the gravitational force exerted by heavenly bodies against each other. Einstein, on the other hand, used the speculative method. In this method, Einstein based his theory of relativity regarding planetary motion on intuition and not by observable facts. In other words, by speculation and intuition, Einstein was able to deduce the conceptual framework of the amount of force exerted by planets as they revolve around the sun. He then reduced this conceptual framework to a mathematical equa-

---

[93] Cornelius Lanczos, *Albert Einstein and the Cosmic World Order*, 112.

tion. This method is not new. The ancient philosophers had used this method, but whereas they failed (as in the case with Aristotle who erroneously believed that the heart is the center of the nervous system), Einstein succeeded. Einstein's biographer, Cornelius Lanczos, commented about Einstein's method:

> The magical thing about Einstein was that he succeeded where the ancients failed. He had the mighty tools of mathematics at his disposal, developed in a slow evolution of more than two thousand years since the time of Plato and Aristotle. The equations of Einstein, arrived at by purely speculative means, told us that indeed the planets had to move around the sun in ellipses and not in something else because our universe is not any mathematical universe, but a marvelously reasonable and well-ordered mathematical universe, pervaded by a supreme Cosmic Wisdom.[94]

Einstein believed that if most of us could not comprehend the Universal Law it wasn't because this Law was not logical but because the intellect of the majority of humans, at this stage of their evolution, had not yet developed to the point of understanding this Law. Only a few humans have a well-developed intellect. But, given time we will get there. Because to understand the universe, all it takes is to understand the laws that operate its components and the relationship among these components or foundations, namely, matter/energy, space, time, and the forces of gravity and electro-

---

[94] Ibid., 12.

magnetism. In his book, *Our Cosmic Habitat*, Martin Rees wrote:

> The structures erected on these foundations are still disjoint. Until there is a unified theory of the forces governing both cosmos (forces) and microworld (matter), we won't be able to understand the fundamental features of our universe; these features were imprinted on it at the very beginning.[95]

Einstein was working on a unified field theory in the latter part of his life. Unfortunately, he died before he could complete this study. Someday, a scientist will pick up where Einstein left off and be able to formulate the equation which governs the operating law of the foundations of the universe. Once science has accomplished this, we are on our way to realizing our full potential as spiritual beings.

Yes, we can make it. As spiritual beings we have the potential to be bona fide members of the chain of being or the community of gods. We can score a victory over our ape heritage, shed our animal beginnings, and superimpose the godly essence on our natures. This is the natural path to follow, but before we can do this, we need to pass the final judgment of our creators--the Sumerian gods. How do we achieve this? We can accomplish this by following the righteous path or by living a moral life.

All beings, gods and humans, are subject to the Moral Law. Einstein calls this law the "moral imperative." In his book, *Out of My Later Years*, Einstein said that "the moral

---

[95] Martin Rees, *Our Cosmic Habitat*, xv.

imperative is the most precious traditional possession of all mankind."[96]

It is the most precious possession because it is the key to the survival of all spiritual beings, including gods and humans. The leading Sumerian gods, except Enlil, were moral and ethical in their behaviour. Immoral behaviour in both gods and men were dealt with accordingly, sometimes severely. The humans who were made to suffer punishment were:

> People who walking in transgression
> reached out with high hand
> Who transgress the established norms,
> violate contracts,
> Who looked with favor on the places of
> evil
> Who substituted a small weight for a
> large weight,
> Who substituted a small measure for a
> large measure.[97]

Likewise, when the god Enlil raped the goddess Ninlil, he was severely reprimanded by the Council of Gods. The following is a passage condemning Enlil's immoral act:

> Enlil walks about in the Kiur (Ninlil's
> private shrine)
> As Enlil walks about in the Kiur
> The great gods, the fifty of them,
> The fate-decreeing gods, the seven of
> them,

---

[96] Albert Einstein, *Out of My Later Years*, 24.
[97] Noah Kramer, *The Sumerians*, 125.

> Seized Enlil in the Kiur, saying:
> "Enlil, immoral one, get you out of
> the city,
> Nunamnir (an epithet of Enlil), im-
> moral one, get you out of the
> city.[98]

Due to Enlil's ignorance or lack of information, he didn't know that when the first viable humans, Adapa and Titi (the Biblical Adam and Eve), were created, they carried in their spirit the great god Enki's essence. As one recalls, Adapa and Titi were made from Enki's sperm and a primitive worker's egg. Enlil couldn't have known the finer workings of a great god's essence since he was not a scientist. He was a warrior. Granted that he succeeded in destroying the whole human race in the Flood, humankind will still lived on because what he destroyed was only the physical form. Man's spirit, being pure energy or conscious energy, was indestructible and could not be congealed into matter.

When Enki, with Ninhursag's assistance, created Adapa and Titi, he equipped their spirit or soul with the Moral Sense. This sense is the knowledge between right and wrong or good and evil.

The Moral Sense, embedded in the human mind, embodies the Moral Law. The Moral Law is autonomous and independent of religion and all other external teachings. In his book, *Moral Laws*, Edgar Brightman wrote about the autonomy of the Moral Law:

> Moral law is autonomous and independent of religion and of the exist-

[98] Ibid., 147.

ence of God so far as the obligatory nature of its principles is concerned.

> Moral law embodies the eternal principles of space relations. This does not imply that any human being has an adequate knowledge of the one or the other; but it does imply that both are principles of reason which may be discovered by autonomous thought.[99]

When the German philosopher Immanuel Kant said, "What you ought, you can," he was, of course, referring to the obligatory nature of the Moral Law. Every human being is obliged to execute this Law in all her/his actions because it is her/his nature to do so.

In a situation involving moral judgment, the Moral Sense springs into action--it mobilizes the Moral Law, and controls the mind and tells it the right and wrong action to take. The following is a passage showing how a mind, controlled by the Moral Sense behaves:

> The mind . . . perceives the truth, as the eye doth light . . . the mind perceives that white is not black, that a circle is not a triangle, that three are more than two and equal to one and two. Such kind of truths the mind perceives at the first sight of the ideas together, by bare intuition, without the intervention of any other idea, and this kind of knowledge is the clearest and most certain that human

---

[99] Edgar Brightman, *Moral Laws*, 269.

frailty is capable of.[100]

Adherence to the Moral law leads to our survival; defiance leads to our destruction.

Doing the right thing or upholding the Moral Law is considered the most virtuous act since it is the true nature of a spiritual being. An individual who upholds the Moral Law is usually blessed by the gods and enjoys good health and prosperity. A classic case is Ziusudra (Noah in the Bible). Although it took a lot of time and effort on the part of Enki to persuade Enlil to spare Ziusudra because he was a righteous man, Enlil, initially, refused to oblige Enki's request for reasons of his own. In the end, grudgingly, Enlil consented to abide by the Council's decision to allow Ziusudra to continue to live. The Council voted to let Ziusudra to continue to exist because he was a moral man--he upheld the Moral Law. On the other hand, an individual or society that lacks moral direction destroys itself automatically. We see this phenomenon illustrated in most literatures of the world down through the ages. An example of this is the Biblical story about the destruction of Sodom and Gomorrha. Lot asked the angels if they would spare the twin cities if he could find at least one righteous man. The angels replied they would. Unfortunately, there wasn't a single righteous person. So the cities and all the humans in them had to be destroyed. The humans were destroyed because they had rendered themselves undesirable specimens. They had regressed into the animal state. By regressing into the ape or animal state, they had allowed the subjugation of their spirit (the god's essence) by the beast from whom they were fashioned. Since the human form was no longer suitable to house the godly or divine spirit, it had to be destroyed to release the spirit.

---

[100] James Bonar, *Moral Sense*, 20.

Likewise, in post Biblical era, we see the market inundated with literature dealing on the theme of good and evil, where good is adherence to the Moral Law and evil is defiance to it. Usually, the theme of evil is portrayed in the protagonist's decision to break the Moral Law in favor of the ephemeral and transitory values of fame and fortune. Perhaps, the best literary pieces which illustrate the pursuit of these values are William Shakespeare's play, *Macbeth* and Fyodor Dostoyevsky's novel, *Crime and Punishment.*

Evil in *Macbeth* is portrayed in Macbeth's obsessive desire for fame. He wants to be king of Scotland. This play is set in Scotland. Macbeth, a general in King Duncan's army, aspires to be king of Scotland. But he isn't sure how to achieve this objective. So, he consults the three witches who foretell the future. They tell him that he will rise from his present position to a position of prominence. Macbeth tells his wife about the witches' prophecy, and Lady Macbeth, emboldened by the witches' prediction, spurs her husband to realize his aspiration. At first, Macbeth is hesitant to kill King Duncan because he has no compelling reason to do so, but in the end, he gives way to his lust for fame and glory. Together, Macbeth and Lady Macbeth murder Duncan when he spends the night at their manor. But before they can initiate their plan of taking over the throne, Macbeth is killed by King Duncan's loyal people, and Lady Macbeth is driven to insanity and commits suicide.

Crime and Punishment is a story about Raskolnikov, a poor young man who wants to be a physician. As the novel opens, he is already in medical school. He can't concentrate or focus on his studies because of financial problems. He lacks the money to provide for his personal needs and expenses involved in his medical program. The money he makes tutoring pupils in the grade school is not sufficient to cover all his expenses. He wants a fortune, a lot of money. Being of superior intelligence, he is able to find a way to solve this predicament. He decides to kill a pawnbroker and steal her money. He feels justified in murdering her because, he thinks she is a liability to the community--she charges exorbitant interest rates and takes the collateral if the pawned item is not redeemed in due time. In short, in Raskolnikov's mind, she is a menace to society and society is better off with her dead. Also, Raskolnikov reasons that with her money he will be able to finish his medical training and help the community. Raskolnikov carries out his plan successfully and commits the perfect crime--he leaves no clues. Without a clue, the police are paralyzed and the murder remains unsolved. Raskolnikov thinks he has gotten away with it. For a while, he does. But as time passes, he finds that he can't sleep soundly as he used to. He is having nightmares. He keeps seeing the old pawnbroker's face with her eyes filled with hate, and hears her voice yelling at him: "murderer." The nightmare keeps recurring night after night until it reaches a point when he can't take it anymore. He must confess or break down. He chooses to confess. He is sentenced to twenty years of hard labor in Siberia, but on the night following his confession, the nightmare stops.

Both stories have the same theme: defiance of the Moral Law. The first, lust for fame; the second, inordinate love for money. Both writers, William Shakespeare and Fyodor Dostoyevsky, came from two different cultures and lived at different times. Shakespeare came from England, Dostoyevsky came from the former USSR; the former lived sometime in the 1500s, the latter in the 1900s. Yet, both writers share the same understanding of the Moral Law, like everyone else, because every human is subject to it.

Shakespeare and Dostoyevsky show, in their respective stories, what happens to individuals when they reject the promptings of the Moral Sense and disobey the Moral Law--

they are punished. In the play *Macbeth*, the protagonist, Macbeth, is killed by King Duncan's loyal people, Lady Macbeth loses her mind and commits suicide. In the novel *Crime and Punishment*, the author shows that the Moral Sense is the thing that cannot be fooled. The Moral Law executes itself, regardless. This is clearly depicted in Raskolnikov. He fools everyone, including the police, because he has committed the perfect crime. He left no clues to his act, and without any clues the murder cannot be traced to him. He thinks he has gotten away with it. But he is wrong. His conscience gets him. The Moral Law executes itself in the form of a nightmare. Until the Moral law is satisfied, his conscience won't let him be. For as long as his crime goes unpunished, the nightmare recurs. Only when he confesses and his crime is punished, does the nightmare go away, for the Moral Sense has been vindicated.

The Moral Sense is the most demanding of all our senses. It demands that it be satisfied before an individual is put in a state of equilibrium. Following the commitment of an immoral act, the person is thrown into a state of imbalance and experiences mental anguish and emotional disorders, sometimes right away, sometimes later. Only when the evil or wrong act is rectified will the person be restored to balance and experience the feeling of having peace of mind, as in the case with Raskolnikov.

Enki was fully aware of what he was doing when he equipped Adapa and Titi (or Adam and Eve, if you will) with the Moral Sense. He wanted humanity to have an opportunity to join him in the community or abode of the gods. Humans can avail themselves of this opportunity by upholding the Moral Law which is embodied in the Moral Sense. Upholding the Moral Law is tantamount to following the righteous path which leads to the abode of the gods. The best among us, Einstein, Kant, Shakespeare, Dostoyevsky, etc., knew this only too well. It was for this reason that throughout their lives they admonished humanity, in their writings, to uphold the Moral Law. Because if we defy this Law, we are on our way to perdition or destruction. Human history is filled with such events. Classic examples are the accounts in the Bible about Noah's Flood, and Sodom and Gomorrha.

Much as Enki wants us to be with him in the abode of the gods, he cannot bend the Moral Law in our favor. He, himself, a great god, is subject to the Moral Law. All spiritual beings, gods and humans, are not above the Moral Law.

Humanity chooses its fate. What will it be?

Babylonian Legends of the Creation, British Museum, Department of Egyptian and Assyrian Antiquities, 1921.

Barlow, Nora. *The Autobiography of Charles Darwin 1809-1882*, London: Collins Clear Type Press, 1958.

Benito, Carlos. *Enki and the World Order*, University of Pennsylvania Dissertation, 1969.

Bergson, Henri. *Creative Evolution*, trans. Arthur Mitchell, New York: Dover Publications Inc., 1998.

____________. *Matter and Memory*, trans. N.M. Paul and W.S. Palmer, New York: Zone Books, 1991.

Bibby, Cyril. *Scientist Extraordinary: The Life and Scientific Work of Thomas Huxley 1825-1895*, Oxford: Pergamon Press, Ltd. 1972.

Browne, Janet. *Charles Darwin: The Power of Place*, New York: Alfred A. Knopf, 2002.

Burkhard, Frederick and Duncan Porter. *Correspondence of Charles Darwin*, United Kingdom: Cambridge University Press, 1997.

Cottrell, Leonard. *The Quest For Sumer*, New York: G.P. Putnam's Sons, 1965.

Darwin, Charles. *Descent of Man, The*, New York: Prometheus Books, 1998.

____________. *Origin of Species, The*, Oxford: Oxford University Press, 1996.

______________. *Variation of Animals and Plants Under Domestication, The*. New York: D. Appleton and Company, 1896.

Darwin, Francis. *The Life and Letters of Charles Darwin*, London: Murray Inc., 1887.

Darwin, ed. Philip Appleman, New York: W.W. Norton & Company Inc., 1979.

Einstein, Albert. *Autobiographical Notes*, translated and edited by Paul Arthur Schlipp, Chicago: Open Court Publishing Company, 1979.

______________. *Out Of My Later Years*, New York: The Philosophical Library Inc., 1950.

Ellegard, Alvar. *Darwin and the General Reader*, Chicago: The University of Chicago Press, 1990.

Encyclopedia Britannica Inc., 2001.

Enuma Elish: The Seven Tablets of Creation, ed. L.W. King, London: Luzac & Co., 1902.

Hull, David. *Darwin and His Critics*, Chicago: The University of Chicago Press, 1973.

Huxley, Leonard. *The Life and Letters of Thomas Huxley*, 2 vols. London: Mcmillan, 1900.

______________. *Evidence As To Man's Place In Nature*, London: 1863.

Jastrow, Robert. *The Essential Darwin*, Boston: Little Brown and Company, 1984.

Kramer, Samuel N. *History Begins at Sumer*, Pennsylvania: University of Pennsylvania Press, 1981.

_______________. *The Sumerians*, Chicago: The University of Chicago Press, 1963.

_______________. *Sumerian Mythology*, New York: Harper & Brothers, 1961.

Lambert, David. *The Field Guide to Early Man*, New York: Facts On File Inc., 1987.

Lanczos, Cornelius. *Albert Einstein and the Cosmic World Order*, John Wiley & Sons Inc., 1965.

Leakey, Richard E. *The Making of Mankind*, New York: E.P. Dutton, 1981.

_______________. *The Origin of Humankind*, New York: BasicBooks, 1994.

Lewin, Roger. *Human Evolution*, Massachusetts: Blackwell Science Inc., 1999.

___________. *Scientific American Library*, New York, 1993.

Longford, Elizabeth. *Victoria R.I.*, London: Weidenfeld & Nicolson, 1964.

Maslow, Abraham H. *Toward a Psychology of Being*, third ed., New York: John Wiley & Sons Inc., 1999.

Milner, Richard. *Charles Darwin: Evolution of a Naturalist*, New York: Facts On File Inc., 1994.

Montgomery, Ruth. *The World Before*, California: Fountain Valley Inc., 1984.

Rees, Martin. *Our Cosmic Habitat*, Princeton University Press, 2001.

Sitchin, Zecharia. *The Lost Book f Enki*, Vermont: Bear & Company, 2002.

______________. *The 12th Planet*, USA: Avon Books, 1978.

Stoczkowski, Wiktor. *Explaining Human Origins*, United Kingdom: Cambridge University Press, 2002.

Sumerian Gods and Their Representations, eds. I.L. Finkel and M.J. Geller, Groningen: STYX Publications, 1997.

Tobias, Phillip. *Hominid Evolution*, Alan R. Liss Inc., 1985.

Understanding Cosmology, compiled by Sandy Fritz, Scientific American Inc., 2002.

Webster's New Collegiate Dictionary, eds. Henry B. Woolf and Edward Artin, Massachusetts: G&C Co., 1981.

Wohlstein, Herman. *The Sky God An/Anu*, New York: Paul A. Stroock, 1976.

Made in the USA
Monee, IL
04 November 2026

67381735R10079